T0275930

KOSSOH TOWN BOY

A PHOTOGRAPH TAKEN IN 1919

Standing, left to right: myself, father, Arthur. *Sitting:* granny
Cole, mother, granny Smart (with Eric). *On the ground:*
Phoebe, Irene.

ROBERT WELLESLEY COLE

Kossoh Town Boy

WITH ILLUSTRATIONS BY

FELIX COBBSON

CAMBRIDGE UNIVERSITY PRESS

CAMBRIDGE

LONDON · IBADAN · ACCRA

CAMBRIDGE UNIVERSITY PRESS
Cambridge, New York, Melbourne, Madrid, Cape Town, Singapore,
São Paulo, Delhi, Dubai, Tokyo, Mexico City

Cambridge University Press
The Edinburgh Building, Cambridge CB2 8RU, UK

Published in the United States of America by Cambridge University Press, New York

www.cambridge.org
Information on this title: www.cambridge.org/9780521046862

First published 1960
Reprinted 1970 (twice) 1971 1976 (twice) 1977 (twice)
Re-issued 2010

A catalogue record for this publication is available from the British Library

ISBN 978-0-521-04686-2 Paperback

*Dedicated
to the memory of my
Mother*

CONTENTS

7

LIST OF ILLUSTRATIONS

9

Cocoon in the Sun

THIS is the story of a little boy. In a way it is an auto-biography; but actually the aim is less personal. If I know this boy perhaps better than others know him, it is because, as I look back on the years that have trailed into the unknown, I can see, somehow, a continuing evolution.

I wonder what my father, or my mother, were they alive today, would say of this boy of theirs? Has he grown big in realisation of their dreams? Or has he merely accumulated years? I do not know. But somehow I feel certain that, if I had to live my life over again, I would not have it different.

But this by itself would not be an excuse for this book. No, my real reason is a sneaking feeling that this boy, in those far-off days, with his rough edges, his foibles and nuances, his grimaces and gaucheries, in the contortions by which he was wrapped in the cocoon of childhood innocence that basked in that tropical sun, somehow represents a pattern. He is one of a group of several ex-little boys I know, who have been not so different after all. And it is really of that group that I wish to speak.

It was a happy group. And when I say 'group' I do not mean a particular collection of boys, nor girls for that matter; but rather the generality of young people who were my companions at play, in school, church, rambles, picnics, and other group activities.

To think of those days brings a glow, a reinduction of that warmth which a whole youthful of tropical sun has burnt deep into my being.

Or was it the warmth of a happy home life? Of parents ideally suited to each other; married young enough but not too young; only a year between them: sufficient to pay due homage to the African view that man is the protector of woman, and small enough to enable them, nay force them, to share their problems in mutual comfort and succour?

As to this son of theirs, he was only very little when his parents let him know that they had prayed to God that their first-born should be a son, and that, should this happen, they would give him over to His service.

There is a strong Old Testament strain among my people.

The first people in West Africa to embrace Christianity *en masse*, with a strong tradition of having been delivered by God rather like the children of Israel under Moses, the Krios of Sierra Leone have for two centuries nursed proudly their connection with Britain. The very name of their capital, Freetown, they had been taught, was a sign of that Purpose which, through God's inspiration, had found expression in the eighteenth century in such philanthropists as William Wilberforce, in whose honour Freetown's Town Hall[1] had been proudly erected, and stood facing northwards across the Atlantic Ocean.

So, when their first-born child turned out to be a son, my parents gratefully promised him to God. Thus from early childhood I was made aware that my life implied a debt of service. The story of Abraham and Isaac, and that of the young Samuel in the Temple, figured prominently in my early instruction.

But for a long time it seemed that God was not keen on the bargain; and in later years my grandmother would remind me forcefully that it had been a struggle to rear me. This

[1] Destroyed by fire October 1959.

information she was wont to put out in moments of domestic stress in which I was involved. This point of view was backed by the family doctor, a product of the Royal Colleges of Edinburgh, and the son of an old Krio family. He considered my continued existence as one of the miracles of his practice, one of those surprises which somehow made medical work rewarding.

Instead of bones, as a baby I seemed to have been made of gristle, '*croun-croun*' in our language, a wonderfully suggestive term. Added to this I had an unusually long neck; and a large head. It was not water on the brain, the doctor explained.

Fortunately, although I was born in a country where malaria and other tropical diseases were the order of the day, diseases from which children of temperate climes are mercifully free, I did on the other hand have the advantage of being born into what in England would be called a middle class home.

In West Africa, you are either white or black. If you are white it does not matter what you are, because in any case you will not be staying in the country for good. You come for a few years, do your job, amass your wealth, win your converts, and go away, leaving the country to the Africans and their mosquitoes, their sunshine, their poverty, and their hopes.

I use the term 'middle class' almost subconsciously, because I have been conditioned by years of living in Britain and Europe, where these things matter. If my family were English, or what until recently used to be officially termed 'British subjects of pure European descent', they would have been upper middle class folk.

My father was a Civil Engineer, a Water Engineer; furthermore he was head of his Department, the first African in the twentieth century to have that distinction, and indeed for almost a generation the only one.

This was important in view of the conditions existing at the time. In the nineteenth century a number of Africans held superior posts in the public life of British West Africa. For example in 1864 a Sierra Leone Krio, Samuel Adjaye Crowther, had been consecrated bishop of the Niger Diocese, the first coloured bishop in the whole world since the Reformation.

In 1859 two Sierra Leone (Krio) doctors qualified in Britain and were commissioned in the British Army and served in West Africa. Between 1885 and 1895, the Chief Medical Officer of the Gold Coast was another African doctor, also a Sierra Leone Krio.

But at the turn of the century this policy was reversed, and for the next forty years Africans were relegated to junior posts in the public service. One of the first moves was the creation of a unified West African Medical Service in 1902, restricted to British subjects of pure European blood. Henceforth African doctors, even the most senior, were relegated to separate junior appointments as 'African Medical Officers'.

In those prevailing circumstances the appointment of my father in 1905 to the post of Superintendent of the newly established Freetown Water Works, in succession to an Englishman, was unique.

His salary on retirement after almost forty years' service was £360 per annum. If any further proof were needed that he belonged rightly to the middle class, this ratio of moderate salary to large responsibilities should suffice. But even so he was much better off than many a fellow African, and many a colleague who had shared with him the glorious days of life at the C.M.S. Grammar School, Freetown.

As to myself, my greatest pride as a boy was that people would often stop me in the street and say: 'You are the living image of your father!' and proceed to tell me what a brilliant student he had been, and what a good man he was. I need hardly say that this sometimes had its drawbacks, especially on those days when boys would be boys, and I would come home only to find that some interfering person had stopped my father in the street and said, 'I saw your boy today, and he was doing so and so'.

Usually the 'so and so' was the very last thing I should have wished my father to know about.

I can remember the only occasion I ever played truant. It wasn't exactly truancy, but rather truancy in reverse. I was late for school; and it wasn't the first time, nor the second, by a long run. This was despite the meticulous efforts of my parents to get me away in good time for school. But I was made that way. Anyway this day, when I arrived in class some quarter of an hour late, teacher sent me home.

I dared not for the life of me go home. Instead I decided to while away the time until the usual hour for getting home from school arrived. I never spent a more miserable time. Never did I realise it was so difficult to put in those hours from 8.30 a.m. to 3.0 p.m. I was thoroughly fed up and sick and tired when eventually I crawled home at the usual time.

'Have you had a good day at school, dear?' mother asked.

'Yes, 'ma.'

'What did you do?'

'Arithmetic, reading, geography, and moral instruction, 'ma.'

Moral instruction was one of the subjects taught. Mother did not actually kiss me; African mothers did not kiss their children. But she could not have been more motherly in the way she bustled my brother and me to have our tea; for she saw we were tired out.

When, however, father came home he was fuming. Needless to say he had heard of the episode not only from teacher, but from at least three other people who had passed along Circular Road, and had seen me playing on the porch of Tabernacle Church. I should like to say that I had the caning of my life; but in my short life I had so many canings that altogether it was just one of those days!

But long before that incident I had passed through a struggle for survival. As an infant I was always ailing, rather a puny weakling; and I am sure that, but for the advantage of enlightened parents and a good home, I should probably have passed away. Very many children died of pneumonia and fits, especially in the rainy season.

My head and neck nearly strangled me more than once while I was a baby. On one occasion when I was asleep I rolled sideways and slipped between the bed and the wall. That is, I almost slipped through. My body did, but my head stuck. But my grandmother calmly drew the bed out and let me drop to the floor.

This was my mother's mother, granny Smart. She was of Ibo stock, and brought to the family the sturdy nonchalance of Eastern Nigeria. My father's parents on the other hand were Ijebus and Egbas from Yoruba-land.

On other occasions, when I was being carried along on my grandmother's back, in the African fashion, my head would loll sideways on its weak long neck. It was a frightening spectacle. It seemed as if the neck would snap with the weight of the head.

The African woman uses her back for a pram. It is comfortable and warm for baby, and convenient for mother.

She straps the child on her back with a *lapa*, a broad woven cotton cloth of bold colours, knotted in front over the breast. Over the *lapa* is tied a narrower wrap or *oja*, so arranged that it fits snugly under baby's bottom, and is straddled by his legs. In this way baby is held firmly against his mummy's back, and prevented from sliding downwards.

This is an ideal position for any normal baby. But in my case this often proved a hazard, as my head would drop backwards over the top of the *lapa*. Passers-by would shout to my grandmother:

'*Mami da pikin i ed de fodon!*'

(Missus, your baby's head is falling off!)

My grandmother for her part, quite unruffled, would reply:

'*Lef am de ya!*'

(He's all right, thank you!)

Granny Smart was very popular with us children. I do not know whether it was a sixth sense, or a seventh, which told us children that she was a pal, even though she could be one of the strictest of disciplinarians. We loved her; all of us without exception.

She lived at the village of Regent, four miles up the mountains from Freetown; and how we loved to see her, whenever she came to stay with us! We could almost sense when she was coming, and we would run to the end of the street to give her a hug. We had to throw our little arms very wide apart to hug her. With this went a big grin and a yell of delight.

She was comfortably built, plump but not too fat. She always had a smile for her grandchildren, yet could give a sharp cuff when the occasion warranted. If we children did wrong she would either ignore it, or else she would chastise us there and then. She had not the habit, which some other

relatives had, of holding our misdeeds over our heads till our
father came home, and then reporting us for the necessary
punishment, a procedure which spoilt the rest of the day for
us while we waited for the inevitable.

⊰ 2 ⊱

Birth of a People

I WAS born at no. 15, Pownall Street, in the Kossoh Town
district of the Eastern Ward of Freetown, capital city of
Sierra Leone. The number was later changed to no. 13. The
day was 11 March, a Monday.

In Freetown all the street names are British. As if to
emphasise this incongruity, nothing could be less British and
more African than the street scenes. Pownall Street lies
practically at the centre of Kossoh Town. Kossoh Town,
Fula Town, Cline Town, Fourah Bay, Mende Town,
Bambara Town, Kru Town, and other such names, are all
districts of Freetown. In most cases the names refer to the
people who were originally settled in each district after
Freetown was founded.

Many of those settlers were free and self-respecting Africans
who in the bad days of the first half of the last century, when
slavery had been banned by the British Parliament but was
still a profitable trade, had been tricked by their own kins-
men and sold to the white slavers. They had, however, been
rescued by British warships on the high seas and taken to
Freetown, where they became free once more and were given

British citizenship, education, a new religion, and new life. They were termed 'Liberated Africans'. They should have been termed 'Reborn Africans'.

Between 1807 when the slave trade was outlawed and 1863 when the last slaving ship was captured some 50,000 Africans were released and set free in Freetown. Many returned home, but many also remained in Freetown, where they formed a polyglot community, hailing from every part of the African coast, from Cape Verde to the Congo, and from as far inland as Timbuktu, and Sokoto in Northern Nigeria.

In addition to the 'Liberated Africans' there were the Maroons, or freed slaves from Jamaica, the Nova Scotians, or remnants of American Negro slaves who had fought on the side of the British in the American war of Independence, and a few of the descendants of the original band of two hundred freed slaves who had gone out with white 'wives' from the streets of London, in 1787, under the auspices of the British philanthropist, Granville Sharp, to found the new colony of Freetown. In addition there were a number of primitive natives who technically had not been exported as 'slaves' nor 'liberated', but who were drawn to the new town from the hinterland and from neighbouring countries.

A sample study of the voices heard in the streets of Freetown in 1849 showed no less than ninety languages and dialects. From this motley they rapidly settled down to become the people now known as Krios, with distinctive customs and a language of their own. Any Kossohs in Kossoh Town had long cast off their original tribalism and become Krios by the time I was born. Krio, pronounced 'Cree-oh', is African for Creole.

This transformation took place in less than a single century. But so definite and striking were the people so born, so

'British' were the Krios in outlook and sympathy, that Queen Victoria called their country her 'Ancient and Loyal Sierra Leone'.

★   ★   ★

As to myself I was given at birth the Christian names Robert Benjamin Wellesley Ageh. At school I was promptly nicknamed Bob 'British West Africa' Cole, from my initials. Of all these names the only African one is Ageh; and for this I have to thank my great-grandfather. Curiously enough we have to thank him too for the most outstandingly British of our names—'Wellesley'.

Both 'Ageh' and 'Wellesley' are shared by the male members of my family. My father (born in Freetown, 22 December 1880, died 19 June 1943) was called Wilfred Sydney Wellesley Ageh Cole. My grandfather (born Kingston, Jamaica, 6 August 1857, died Freetown, 23 February 1895), was called Augustus Benjamin Wellesley Cole: 'Augustus' after his birth month; 'Benjamin' in the biblical tradition of Jacob, because he was the youngest of a long line of children.

I do not know how my great-grandfather Ageh managed to be in Kingston, Jamaica. Most probably he went as one of the Liberated Africans, who, after slavery was abolished, opted to go to the West Indies from Freetown, under the apprenticeship scheme which was later stopped in 1840.[1] We know that he was of Egba (Yoruba) stock. In Jamaica he is said to have married a 'West Indian' woman, that is a native of the West Indies as distinct from himself, a foreigner, although they were both of African origin. When his wife died, he decided to return home to Freetown; but he was

[1] See Colonial Office Circular Despatch to the Governor of Sierra Leone, 1840.

only 'allowed to take away' his youngest child, that is grandfather Augustus, who was then four. This too is significant of the times.

Of the circumstances in which he originally left his native Egba-land in Yoruba Nigeria, as a young man, we do not know. But we can hazard a guess. Trickery, kidnapping, one of the many tribal wars, possible religious persecution—any or many of these factors might have played a part in an unwilling exodus. On the other hand he might just possibly have gone willingly, attracted by the lure of the new civilisation which had been established in Freetown—the first in tropical Africa.

What we do know is that he was a Christian convert. Hence the surname, Cole. It was the practice in those days for African converts to give up their native names and other 'marks of heathenism', and take the name of a good Christian Britisher, usually the missionary who converted them. There are so many Coles in Freetown and Lagos that clearly there must have been a very hard-working missionary of that name in the early part of the last century.

But whatever the circumstances of our grandfather's translation or transmigration, he was of birth and importance in his tribe, the son of a ruling house, a man of strong character, and a noted warrior. While other converts chose the names of local white men, nothing but the most illustrious soldier of the day in the whole wide world would suit him.

So he adopted the family name of the Duke of Wellington. I do not know of any other family in Africa called Wellesley. Incidentally by virtue of this name it is possible to trace the period early in the nineteenth century when this ancestor embraced the new Christian civilisation, and so started our Krio family. As to 'Ageh', we are told it means 'man of iron'.

My ancestors seemed to have derived great pleasure from their names, and to have passed them on from one generation to another. Thus my father's three sisters all had 'Sally' as their second name after their mother, who was baptised Phoebe Sally Vivour Pratt (b. Freetown, 18 October 1848, d. 18 June 1941). This lady (the 'Granny Cole' of this story) was herself named Sally after her own mother, who on baptism had given up her native Jollof name for Sally. The Jollofs are a people who inhabit Senegambia.

This same great-grandmother, the original Sally, was married twice. Her first husband was a Mr Vivour, a Kossoh man. When she married later a Mr William Pratt, an Ijebu (Yoruba) man, she added the Vivour, and her children by the second marriage were Vivour Pratts. The Kossohs are a Sierra Leone tribe. The Ijebus are Nigerian.

Thus we Krios have many roots.

*　　*　　*

Those years of the early and mid-nineteenth century must have been full of romance for our ancestors, despite the insecurity which no doubt abounded. I have referred to my father's paternal grandfather who had found himself, though a free man, in the West Indies. The same thing happened to father's maternal grandfather, the William Pratt who has already been mentioned. He, on his return to Freetown, could not settle down, but emigrated to the rich volcanic island of Fernando Po off the coast of Nigeria and the Cameroons, where skilled craftsmen were in demand. He had learnt the trade of a master carpenter in the West Indies.

At Fernando Po he noted the similarity of the rich soil with that of the West Indian Islands, and he sent for cocoa seeds and started planting cocoa. He it was who thus introduced the cocoa tree to Fernando Po, and it was from that island that cocoa was introduced to the Gold Coast (Ghana) and Nigeria.

He did very well out of cocoa and became a sort of country squire. He was able to keep his son, my great-uncle Jacob Vivour Pratt, in England for many years at school and later studying law. He was called to the Bar, but on returning home he found he did not like law as a profession; so he returned to the United Kingdom to study medicine at the Royal Colleges of Edinburgh, graduating eventually in November 1895. He was the eighteenth Sierra Leonean to qualify in Britain as a doctor. Incidentally the next year Freetown's leading barrister, a Krio, was created the first African Knight, Sir Samuel Lewis.

My mother's family on the other hand were Ibo, also from Nigeria. Her family name was Okrafo Weekes Smart. Okrafo is Ibo. 'Weekes' was after Bishop Weekes, the English missionary who, before he was consecrated bishop of Sierra Leone, laboured for many years in Iboland. There he was very friendly with a local Chief, who was on the verge of becoming a Christian, and as a preliminary had professed to give up all the old heathen practices, including the destruction of twins at birth. The people believed that twin births meant ill luck for the tribe.

Then one day the Chief's senior wife gave birth, and the missionary called to offer his congratulations and suggest that the child be baptised. To his horror he found that the lady had given birth to twins, and that according to tribal custom the baby princes had been put out to die by exposure.

He acted swiftly, rescued the babies, and took them to the new settlement of Regent, near Freetown. They were baptised and given the name 'Weekes', which they passed on to their children.

24

☙3☙

A Son is Born

THE more I try to recall the earliest impressions of my childhood, the more surprising the result becomes. I can see
myself as a new-born baby, barely a day old, possibly a few
hours old, receiving the vigorous attentions of my maternal
grandmother.

She sits on a stool in the middle of an airy bedroom, a proud
contented look on her face. On the floor beside her is a broad
metal basin, rather more than two feet in diameter, and about
eight inches deep. This is half-filled with soapy warm water.
Across her knees is spread a waterproof sheeting. On this she
supports me with her capable left hand, while with the other
she anoints me liberally from head to feet with a rich lather
of native black soap. My eyes are tightly shut, my mouth
wide open in a yell of protest. My arms and legs are crossed,
and my body still curled up.

As she vigorously rubs away, the yelling continues, and the
roomful of women look on approvingly.

'*Man pikin don cam. Allelujah!*' (Unto us a son is born!),
they cry.

On a single bed along the inner wall my mother lies, quiet,
a strange look on her face. This is my father's room, converted
into a temporary maternity ward. The connecting door leads
to her own room on the other side of the partition.

Having thoroughly lathered me, my grandmother dips me
into the basin of water, and scoops the contents all over me.
As the water runs into my mouth, I shut up sharp. Half
choking, I gulp hard, swallow, and let out a yell of even

25

louder protest. My grandmother ignores me, and carries on with her task. Then she lifts me on to her lap, and towels me briskly. My body gleams brown like a newly minted penny, the pink flush of dilated blood vessels showing underneath the copper skin.

The assembled ladies take a look at my fat little face with its squashed nose and puffy little eyes, and say admiringly:

'Just like his daddy!'

There is a coal-pot, that is an open charcoal brazier, burning brightly near at hand. From it someone lifts out a small oval stone about six inches long, worn smooth. Normally it is the 'daughter stone' which is used to grind pepper, spices, egusi, lokos, and other cooking ingredients on a large flat mother stone. Now, it has been washed and brought to a red heat, and so thoroughly sterilised. Some water is poured on it, causing a fierce sizzle: but it still remains dull red hot.

A little folded piece of freshly laundered cotton is dipped in a glass of water, squeezed out, and pressed on the stone. It is then lifted up, allowed to cool a bit, and then applied firmly to the end of the birth cord dangling from my navel. Thus sterilised the cord is folded and bandaged into position. This is repeated daily until it drops off.

Finally two pieces of camwood are brought, a daughter and a mother piece. A little water is sprinkled on the latter, and the smaller one rubbed over it briskly in a grinding motion, producing a quantity of the red camwood liquid dye. This is rubbed over my body, which now becomes burnished red. Finally my grandmother picks up a little metal pot of 'kenge', a red strongly perfumed pomade of German make, and richly anoints me. I smell to high heaven, to the rapturous delight of all present.

'The *agbo*, don't forget the *agbo*!' they remind her, as the door opens, and somebody comes in with a small half-gourd full of a herbal infusion. My grandmother tastes it to check the temperature, and then lifting me up gently she pulls her skirt well up her plump left thigh.

She lays me along this limb, with my head over her knee, her left arm supporting my back and head and holding the gourd, some of the contents of which she pours into her right hand held cupped against my mouth. She heaves in a quantity of the concoction smartly as I open my mouth to cry. I choke, swallow instinctively, take a good breath and open wide for a tremendous yell. But she is ready, and immediately I receive another quota into the back of my throat. I splutter, shut my eyes tight, kick out, go rigid with rage, and open wide for a truly nasty yell of protest; but again she is ready and heaves in the right quantity of *agbo* with practised aim. So the duel continues, each yell throttled at birth by a quaff, followed by a swallow, a holding of the breath, and an attempt at an even louder yell, which meets with the same inexorable fate, until the gourd is empty. So I have my first drink in this world.

Then, still not at all amused, I am dressed in a long cool cotton frock, and handed over to my paternal grandmother. She looks long and silently at me, nods several times, and without a word passes out of the room to the parlour, where a number of relatives and neighbours are waiting. There I am shown round proudly to the accompaniment of many words of commendation, mostly biblical. Finally I am taken back to the bedroom and handed over to my mother.

We often hear of the Mona Lisa, whose smile is said to be enigmatic. Have you ever looked at the face of a mother who has just had her first-born son?

*　　*　　*

What I have described actually happened. I saw it myself: the same room, the same low bench, the same grandmother, when my brothers Arthur and Wilfred were born and my sisters Phoebe and Irene.

In reality two years separated each birth from the next. But what are two years in the mind of a little boy? Or in the memory of later life?

In my mind those events have been telescoped into a symbol, a symbol repeated on each occasion, until it did not matter who the central figure was. In this drama the principal actor could be any one of us children, including me. I was the first, that is all.

In this picture somehow I do not see either a doctor or a midwife attending at these births, although we were what could be termed private fee-paying patients. What is more, our father was friendly with most of the doctors. But there were few doctors in private practices, not more than three in those days.

The person who 'grannied' us, that is acted as midwife at the birth of myself and my next two brothers, was one Mammy Jebete, I'm told. She was the grandmother of my father's half-sister, our Aunty Dorcas. As such she must have been about the most senior granny in Kossoh Town at the time, and undoubtedly very experienced.

When the doctor said my survival was a miracle he was referring not to my birth, for which he could not claim credit, but to the serious ailments to which I was apparently subject during my early childhood.

Almost the only medical treatment available in those days was that provided at the Connaught Hospital, Freetown's small general hospital. This medical service was limited to treatment at the out-patients' department of the hospital or as an in-patient. Obviously it did not involve a doctor going out to attend a confinement.

There was in addition the Princess Christian Mission Hospital, fondly called the Cottage Hospital by the people of Freetown. This, as its name implies, was a missionary foundation, staffed by British nursing sisters who came out as missionaries.

This hospital and the bishop's residence were in the same grounds. Curiously enough the central point of the bishop's compound was exactly opposite the bottom of Pownall Street. Bishop's Court was part of our parish of St Philip's, and the bishop often worshipped at our church.

In those earliest days of which I am now speaking the bishop of Sierra Leone was Bishop Walmsley, and I have since found out that he used to be vicar of St Aidan's Church, in Nottingham, before he went out to Sierra Leone. A tall, gaunt, typically English gentleman, he was adored by everyone, children and grown-ups alike, men and women, Christians, pagans, and Mohammedans. When, many years afterwards, he died in Freetown, the whole city turned out for his funeral and the streets were jammed from the cathedral to Kissy Road cemetery three miles away.

He knew our family well. Father, who all his life was pastor's warden at St Philip's Church, and was a member of a number of diocesan committees, was a personal friend of his. He visited our home on more than one occasion that I can remember. On his rounds in the city, he usually either walked or rode in a rickshaw, wearing a fawn-coloured flat-topped pith helmet, which somehow did not seem out of place with his frock coat and gaiters. If he was walking and he saw any of us he would stop us for a little chat.

The popularity of the Princess Christian Mission Hospital

was due very largely to the love and respect which Bishop Walmsley inspired in every citizen of Freetown. The out-patients' department was overcrowded. The patients were women and children, including maternity cases.

Thus when I was due to be born, had my parents wished, all my father need have done was to have conveyed my mother down Pownall Street to Fourah Bay Road, turned left along the latter until they came to the gates of the hospital, only a few hundred yards away, and lo! I should have been born at the Princess Christian Mission Hospital, instead of at no. 15, Pownall Street.

If that had been done, it is possible the change might have speeded or delayed the time of my birth. Thus, instead of my being born at 5.25 in the afternoon of Monday the eleventh of March 1907, the time might have been different by a few hours. In that case would I have been a different person? I often wonder, for the following reason.

When I was eleven years old my father visited England on a tour of water works and other municipal installations in various parts of that country. Two mementoes I have of that visit; at least I had until a few years ago. One was a coloured postcard of the Clifton suspension bridge bearing inset the picture of a local Bristol beauty, called Lally Charles, or possibly Lally Babs. Her oval features, creamy-rosy com-plexion and striking beauty made a lasting impression on me. I hid that postcard and kept it for years. It was my first love.

When my father returned home he brought a horoscope of my life which he had had cast by a famous English astrologer. This was based on the exact place and time of my birth, and was so accurate in the way it described my character, especially its weak points, and in its forecast year by year of what would happen to me, including the exact year in which I should leave my homeland for the United Kingdom (actually this happened ten years afterwards), that even now I am amazed when I think of it.

The point, however, is this. As astrologers generally work on the basis of the position of the heavenly bodies in relation to the longitude and latitude and the exact time of the subject's birth, what would have happened if, through moving from our home to the hospital, the time of my birth had been different? Would I have developed a different personality, and would my career have been otherwise? I often wonder.

❦ 4. ❦

My early Infancy

THE earliest impressions of my childhood are sensuous ones. Home to me was a nest, characterised by warmth, cosiness, loving parents, grandparents, aunts, uncles, cousins, foster brothers, the whole held together as one large family, in which there was never a commotion or revolt. A particular corner of the house meant more to me in those earliest days than anywhere else. When I say corner what I really mean is a little portion of one floor; the ground floor.

The room where we were born was upstairs. But this early corner of the nest was downstairs, in my grandmother's quarters; my paternal grandmother that is: our granny Cole. And somehow as I look back on it, the things I can remember occurred at night, when it was dark. It may be because dusk always came suddenly, at half-past six; just as it was dawn suddenly at five in the morning. There is no twilight in Sierra Leone, nor indeed anywhere in tropical Africa.

Our cosy corner was not a big area of the floor, but just a small patch of ground, sufficient for two little boys to sit and play and fall asleep in when tired, and yet leave room for grown-ups to walk round, or sit and talk.

It was in one corner of the main part of the parlour downstairs. In the centre of this area stood a large round mahogany table, between three and four feet across, with carved tripod legs. The top could be swung vertical by releasing a special wooden stop underneath. On this table during the day on a central antimacassar rested the family Bible, the family album of photographs, and one or two special framed photographs. At night these were moved aside, and in their place a large table lamp stood.

In those early days this was a tall kerosene lamp with slender glass shade surrounded by a bright glistening globe. Later a pressure incandescent lamp took its place, to be displaced later still by electric light. The lamp threw a lovely warm shadow around the table, and provided a wonderful umbra for two playmate brothers.

My brother Arthur and I lived in a cosy world of our own, in which the slenderest incident was full of interest. Even just to sit and talk and count the legs of the chairs or of the people sitting at the table, to watch the shapes and shadows cast by the lamp, or the huge moths which would fly in through the wide-open windows and dash against the bright light, was entrancing.

What we were doing downstairs in this part of the house I do not know; for our quarters were upstairs with our father and mother. Before my parents got married my father had added a second story to the large single-storied five-roomed wooden house on stone foundations. There he took his bride to what was in effect a large four-roomed upper flat with an annexe and a separate entrance at the rear.

The timber for this modern house was all imported from England, as was the material for the corrugated metal roof. How we loved to hear the rain pelting down and rumbling on this roof! It drummed a tattoo which soothed and excited us. And when the thunder roared, and the lightning flashed, we would run and hide under our father's bed, not in fear but in a thrill of excitement. The dark always seemed to fascinate and attract us.

I should like to think that we were downstairs in our favourite corner of the parlour because our parents were out for the evening. Probably grandmother was baby-sitting. If so it showed how convenient was the African way of life, each member enjoying their independence and yet near at hand ready to lend a helping hand.

It was in this corner in the parlour downstairs also that we used to make our challenge on New Year's Eve.

'Please, mama, may we come to watch-night service with you and papa?' we would beg.

They did not reply that we were too young to go to this midnight service in Church. Instead they would say to us:

'All right, if you are awake, you can come to the service.'

And so we would dress up long before seven o'clock and sit in our corner, and manfully keep awake for half an hour. Hours afterwards, when our parents returned from service, they would lift us up gently and take us upstairs to our proper sleeping quarters.

It was in this cosy corner also that my brother and I would fall out, as children do. I cannot remember any specific causes for these differences. What I do remember is that invariably when the case was taken to father, Arthur would win and I would lose. He was always able somehow to marshal his case while I was usually so incoherent with righteous indignation that I was unable to state my case clearly.

Perhaps it is no wonder that today Arthur is a lawyer. I have only once heard him plead, and I should not like to have to challenge him in court. Soft-spoken, suave, persuasive, he is able in his first few sentences to catch the sympathy of the judge and hold it to the end. Certainly he always did with our father, who was himself an impartial, if strict man.

From as early as I can remember our father always taught us to think for ourselves, to give reasons for what we did, and to develop judgement. Above all to be fair.

Did we want a new suit? He would ask us why. Of course we had sufficient sense not to admit it was because our friends had a new suit. Or a new pair of shoes for that matter.

'Well, why do you want a new suit?' he would insist.

'Because it would be nice to have a new suit, sir.' We always said 'sir' to him, and 'ma' when we spoke to mother.

'Don't you understand that suits cost money?' he would ask.

'Yes, sir.'

'And that to have money you have to work hard for it?'

'Yes, sir', we would agree quite readily. For the moral that he was pointing was obviously that on no account should we waste things or do anything which involved waste; because in the end everything cost money or labour or some sacrifice. Therefore if we wanted a new suit, seeing that we already had some, there should be a special reason. He would help us.

'Is the suit too tight for you?' he would ask. 'Or is it torn?' It was never torn, for any tear was promptly mended

2-2

by our mother, who was a famous seamstress. In the end we would get our suit, either at Christmas or at Easter or on our birthdays.

But all along the line our father helped us to develop confidence. He wanted us to look on our parents as friends. No African child would consider his parent his equal, at least not in our days. But we certainly looked upon our own parents as our friends and we were not frightened of them, for we knew they were on our side. As to respect they had that one hundred per cent from us.

While the cosy evenings are associated with our grand-mother's quarters downstairs, in my memory I seem to have spent the hours of daylight upstairs with mother. Here during the day, when father had gone off to work and the foster brothers and sisters to school, leaving the servants and other members whose duties kept them at home, I followed my mother from room to room, and no doubt got in her way.

I have two marks which confirm that I did get in her way. One is the middle finger of my right hand, the nail bed of which is crushed and scarred. One day I was standing beside my mother as she sewed, when, fascinated by the flashing needle of the sewing machine, I thrust my finger at it, and it was promptly transfixed by the shining dart.

On another occasion I was helping my mother. She was downstairs in the kitchen across the yard, and I was fetching a large bottle of red palm oil for her. I was the type of child whose mind was always wandering and eyes roving. I do not know what happened on this occasion, but the next thing was that I came tumbling down the steep staircase, and rolled right down to the ground floor, even getting past the right-angle turn near the bottom of the stairs. The bottle was smashed and the palm oil bespattered the whole place. I yelled, more in fright than from actual injury. I soon quietened down however, only to burst out in real terror this time, when my grandmother came and, seeing the blood flowing from my punctured thigh, shouted:

'Lord have mercy, the boy is bleeding to death!'

I don't think either 'bleeding' or 'death' meant anything to me, but the way she said it made me realise that something terrible had happened.

I must have been two or at most three about this time. But I remember the contents of the house, the yard, kitchen and outhouses, the hens in the backyard, the dogs, and the cats. We had a particular dog named Fido, who was the nicest dog I have ever known, a big friendly mongrel, with kindly eyes and brown patches on its white coat.

There were other things. The berry tree which grew in the centre of the yard, its spiky thorns, its berries, each the size of large grapes, sweet and juicy. Nearby was the old well, sealed up even before my time. The coconut tree which grew at the bottom of the yard. I remember the first time a coconut fell, and the yell with which an agonised aunt warned me not to go near that part of the grounds.

But one of my greatest thrills at this cocoon stage was the water tap upstairs on our floor, at the landing. What a thrill it was to see the magic of the water spouting out when the grown-ups did something to the tap! I would watch them, and when they had gone would try and do what they had done. Needless to say I did succeed one day. It took the rest of the day to mop up the havoc wrought on both floors.

I am sure my guardian angel must have been working overtime in those days. Certainly on one occasion but for him this book might never have been written. At the top of the stairs a wicket gate was fixed which was always kept shut and on the latch. At first this was the same height as myself; but then, while the gate remained stationary, I continued to grow, until the time came when I was able to climb it. I reached the top, got my foot over and across to—a void! There it dangled for a second, as I wobbled woefully, and was on the point of tumbling headlong, possibly breaking my neck in the process, when my mother dashed up and grabbed me.

Fortunately she did not scold me, as this might have made me want to try and do it again. Instead she made sure that the gate was never on the latch when I was alone. If I wanted to get down, it was easier to open it and walk down than to climb over. If I did not think so, but tried to climb it, then at least the unlatched gate would rock and wobble and frighten me off.

❧ 5. ❧

Our Home

OURS was a large wooden house, with corrugated metal roof and stone foundations, raised some three feet from the ground, with a completely enclosed air space underneath. This was called the cellar. It served for ventilation and raised the house above the level of the ground. There were openings in the side covered with wire mesh, and through these we used to gaze into its mysterious dark interior, and weave thrilling fantasies about it.

It was the type of house peculiar to Freetown, and wherever Sierra Leoneans settle along the West Coast. You

mounted a few steps to a porch and entered through the front door into an antechamber, which occupied most of the front of the house, with windows opening on to the street. This was a portion cut off from the main parlour, into which it opened through a wide arch. This double parlour occupied two adjacent sides of the house. The other two sides were taken up by three bedrooms, and a closet or boxroom. All the bedrooms opened into the parlour. The latter was separated by a lattice door from the rear hall, the back door, and back porch. The dining room was an annexe opening out of the back porch.

A staircase led from the rear hall to our own quarters on the next floor, consisting of a similar number of rooms.

The children slept on mats on the floor. The mats were washed regularly, and easily replaced when worn. They were rolled up and put away in the morning, and brought into use at night.

Across the yard from the house were the kitchen and out-houses including the bathroom, or 'wash house' as it was called. At the furthest corner of the yard was the building which contained the toilet and a second bathroom. The toilet consisted of two cubicles erected over a deep pit in the ground.

In addition there was a large backyard partitioned off. Here was the piped water tap standing in the centre of a concave concrete area where the laundry was done. In the centre stood a large polished granite stone on which the clothes were rubbed, pounded and washed. The whole compound was about a third of an acre.

If grandmother's quarters downstairs had for me that warm feeling of cosiness, our own quarters upstairs meant light, interest and a window into the outside world. The windows on the front upstairs looked down into the street and commanded a view stretching from the green slopes of Mount Aureol on the left down to the grounds of Bishop's Court, and a glimpse of the sea, on the right.

But even better than this we lived in a real Jack-and-the-Beanstalk fairyland. For, just outside the window, on the south side of the room, stood a lofty apple tree. What we call apple in West Africa is different from the European variety. The tree grew to a height of thirty or forty feet. The leaves were about six inches long and four inches broad, and the fruit were bell-shaped, white at first, then varying to pink and red when ripe, and very juicy. They hung in clusters, each apple having a hairy tuft at the open end of the bell.

I am sure most boys would have liked to live in a house where, when they looked out of the window in the morning, they could see those lovely clusters of apples actually smiling at them, as if beckoning and saying:

'Hello there, would you like a bite of me?'

And indeed when I grew older I was able to lean out and pick the fruit from some particularly accessible branch, hooking the latter with my father's walking stick.

In the grounds were many different fruit trees. There was a palm tree which at certain seasons a man would come and tap for palm wine. He would come armed with a large hoop made of tough twigs lashed with freshly cut fibres. He would loop this round the tree and himself. Then, leaning against it, he would proceed to climb up the tree, heaving the hoop up step by step. Arrived at the top, he tapped a hole into the tender growing apex of the trunk, and fixed in it a spigot, the end of which he guided into the neck of a large green glass bottle which he tied in position. The sap dripped into this bottle, and was collected each day, as pure sweet palm wine.

At other times he would come to pick the coconuts. This was far more difficult, because this tree was twice the height of the palm tree, fully sixty feet, and also its trunk was quite smooth.

It was always a breathtaking sight to watch him climbing this tall smooth coconut tree, leaning back against the fibre hoop looped round his body and the tree. In his hand he carried a sharp gleaming cutlass with which, arriving at the top, he would attack the clusters of fruit, holding each in his left hand, and separating it with a sharp cut to the tough fibre stalk, then gently dropping the fruit on a soft patch of ground below.

Finally he would scale down the tree. Then the coconut had to be shelled out of its pithy outer coat. He did this by holding the fruit erect with his left hand and hitting it a sharp blow with the cutlass in his right hand, taking care to move his left hand away at the moment of strike. But soon he had all the coconuts shelled, and the fibre was thrown away or else stacked for mixing with firewood.

This fibre is what is used for making coconut fibre matting, whilst the fruit itself is dried into copra, and makes very good oil which is used among other things as a basis for ointments. Coconut oil also makes a highly priced cooking oil. Alternatively the fruit can be grated and the juice extracted and used in cooking rice, to which it imparts a most delicious flavour. This is one of the delicacies for special occasions. The ribs of the leaves are made into brooms and besoms, the leaves themselves are spun into ropes, the petiole of the compound leaf made good cricket bats for us, while the stem makes good cane for fencing or furniture.

This family compound which was my childhood's home was a fairyland which made life a child's paradise. During the day the whole place was bathed in warm sunshine. And at night we had the stars for company. As for the moon, it was so large and bright that one could read by its light alone.

❧ 6 ❧

Mother's Care

THERE were always a number of foster children in our home. They were either relatives whose parents lived in other parts of the country or elsewhere in West Africa, or children of friends who stayed with us in order to attend school in Freetown. There were foster children from my mother's native village of Regent and the neighbouring mountain villages, or else boys, and occasionally a girl, from the indigenous tribes of the interior. These last were illiterate when they came to us, but all were sent to school, and in time they became baptised Christians, and adopted our family name.

Of one such native family five brothers passed successively through our hands as foster brothers. The eldest was Morlai; the others in turn were Yankoo, Shenkoo (who was baptised Philip), Santigi (baptised James), and Kabba (Edward). They were of the Timne tribe, and their mother was a small determined woman, who was ambitious for her children. They all made good. Shenkoo, as Philip S. Cole, moved in later years to Nigeria, and his daughter by a Nigerian mother recently was chosen as Miss Western Nigeria in the Beauty Competitions held in that country.

Home for us was mother. She was the centre of our life,

43

and was never far from us. Even when she was out of the house things were so smoothly run that we did not miss her long.

Each of us had our special duties in the house. Mother was specially anxious that I should not be 'spoilt'—a risk which was real in a household full of servants, cousins, and foster children.

'You never know what may happen to you when you grow up. It is best to be able to look after yourself.'

But life was far from severe. On the contrary we were out playing most of the time, either in our own yard, or outside in the street with other children. There were then no risks from wheeled traffic. There were hardly even any bicycles in the streets.

But although out of doors we had ample room for manœuvring, nevertheless the house was our castle. There we retired at night, and within its comforting walls we sought refuge when we were ill. And we took vigorous action constantly to keep it healthy and free from disease.

When I think of the conditions which existed in those early days, when the only antiseptic available was carbolic, and the only protection against vermin was constant vigilance, I can never be too grateful to our parents for the way they laboured over our health. In such a mixed family group as ours, with children coming from different homes and walks of life, it was not surprising that at times body parasites would be discovered. These were immediately attacked. Head lice were combated by daily baths, regular combing out of hair, and sessions devoted to tracking down and cracking the nits between the nails of both thumbs.

Bed bugs (*chinch*) were a constant challenge; and it was our job to keep them down. We turned up the mattresses and sheets, and squashed the scurrying little beasts. It did us good to see their bloated bodies burst, spewing out blood—our blood. Mother's was a large four-poster iron bedstead decorated with mother-of-pearl. It came originally from Spain, and was a wedding present from Fernando Po. The coiled springs and crevices of the mattress provided ideal hiding places for the bugs. The way we dealt with them was to singe them out with the flame of a candle.

We did this once or twice a week. The other beds were locally made of timber, with morticed cross pieces on which the mattress was laid. These wooden cross pieces were taken out, cleaned, disinfected and replaced. We children loved this job of home sanitary inspectors. The hunt satisfied our sense of adventure. It was of course messy, and the bugs had a smell which was not altogether unpleasant. But they were our deadly enemies and so must die.

Cockroaches were another pest against which we were always on the look-out. They attacked foodstuffs, and left their tell-tale excretions in cupboards, wardrobes and baskets. It was our duty to turn these out and kill the pests. We smashed them with the heels of mother's shoes, wielding these like hammers. When squashed their white mushy entrails splashed out. We were only children, lustful blighters. We revelled in what would nowadays revolt us.

After such messy holocausts we would scrub out the floor and cupboards with disinfectant. At other times we would carry the baskets or drawers gently outside and empty them in the yard. There the cockroaches were promptly pecked up by the hens—a clean, swift and sure fate.

Another pest was mice—they might well have been rats; they seem now to have been so big as I look back. We always kept cats, as well as dogs, and we set traps regularly for the rodents. Now and again one or more of them would be caught. They were promptly taken to the yard and decanted to the cats.

There was never any serious illness in our home. I cannot remember any epidemics or serious contagious diseases like impetigo or ringworm. We did have round-worms, and had to swallow worm cake or powder or Santonin from time to time. But it is a tribute to the standard of domestic and personal hygiene that we did not contract hook worm.

The horror of horrors was castor oil night. This came round about once a fortnight. My brother and I could not swallow this stuff, even though father tried to camouflage it with the juice of an orange or brandy. It floated on top and always clung to our lips for the rest of the day. In the end father used to resort to the cane or the threat of thrashing before we would submit to the torture. I have turned against the taste of brandy and similar alcohols ever since.

We often had coughs. This caused father serious concern. Our coughs would last a long time despite expectorant mixtures from the doctor. Eventually the point would be reached when father would get us to cough into water placed in the chamber, to see if the phlegm floated or sank. If it sank it was lung tissue, and this could only mean one thing, tuberculosis. Fortunately it never sank.

Father had a fine library. In addition to English books on general literature, he had many books from America. These latter specialised in giving detailed advice on the modern way to bring up children, and were strong on the psychological approach to the problem. Looking back I think in the end father compromised between the modern approach and the good old-fashioned ways.

He followed the American methods in home cures, and for

years was a correspondent of the Battle Creek Sanatorium, Michigan, and Bernarr Macfadden. He believed in physical exercise (he was a disciple of Müller), the steam bath, and the enema. To this may be added dieting in times of illness.

We often had fever, which usually lasted about three days, and then subsided with profuse sweating. The usual treatment for this was the American version of the turkish bath. We had two apparatuses; one was a steam couch, the other a steam tent. The former was a couch covered with wicker cane, and its sides boxed in. Underneath were four steam kettles each mounted on a spirit stove. A cage fitted over this couch served as a frame for blankets.

You lay on the couch, the blankets were draped over you and the couch, and the steam kettles were lit. In a few minutes you were in a steaming cauldron, with a strong but not unpleasant smell of methylated spirits. This was continued between twenty minutes and half an hour. It was not long before we felt like boiled lobsters (African variety). Within ten minutes we were usually yelling to be released.

The relief when it came was usually very pleasant. We were plunged into a bath of cold water; a five-foot modern bath. We soaked for a little while and then we had the abdominal version of a sitz bath, that is as we lay we rubbed the lower part of our abdomen to and fro with a cloth. This lasted ten minutes. We were then bundled straight to bed.

This measure usually made us better or aborted a threatened illness.

The other version of the steam apparatus was the steam tent. This was a square canvas box in which was placed a wicker-bottomed upright chair. Under this was placed an ordinary large steam kettle, while another kettle was placed in front between our legs. Ordinary domestic kettles would do, or even pans (basins) of boiling water. The technique was the same.

Instead of the canvas box the same effect can be obtained in any bathroom by simply sitting on a cane-bottomed chair with a kettle of boiling water underneath and another between your legs, and a couple of blankets draping your whole body from head to floor, including the chair and the kettles. You then remove the lids of the kettles for twenty minutes, and then plunge into a cold bath.

During these attacks of fever we swallowed copious hot drinks of teabush or lemon grass, both of which were infusions of delightful fragrance. At the same time we were not allowed palm oil in our 'palava sauce' soups, nor any rich foods.

As to the enema, we were used to this household agency from infancy. A level tablespoonful of salt was added to a pint of warm water. This was run in gently while we lay on our right side, the rectal nozzle lubricated with 'Vaseline'. It was an effective and prompt way of relieving constipation and preventing those ills for which constipation is blamed.

*　　*　　*

My mother was one of those silent and efficient nurses whom nothing deflected from her chosen course of action.

As children we often incurred cuts and scratches on our legs; and these almost always turned septic, despite sundry patent ointments of European origin. Invariably we had to resort to poulticing. This was usually made from starch, freshly prepared and spread thickly on a piece of clean cloth. It was then applied next to the skin. The hotter the better. At least that was what mother believed.

48

One day after my mother had applied a large poultice to a small festering cut in the middle of my left leg, the pain was particularly severe, and I yelled and begged her to remove it. Of course she ignored me, and all that day at school the leg was painful. After the pain of the heat went, a new and peculiar pain took its place. Eventually when the leg was dressed in the evening the mystery was solved. I had a blister as big as the largest tomato I had ever seen; only it was brown.

It took months and several more poultices finally to cure it. The thin ragged scar it left continued to increase as my leg grew. It was exactly at the spot where at football I was likely to receive a kick from an opponent any time he missed the ball. After several such accidents I had to give up football. I was not very good at it however.

We washed our feet several times a day, staining the water red with the colour of the dust. Our feet were long and shapely, our toes supple, the arches springy and elastic.

But at times tragedy would overtake those same toes, when they happened to smash against an unsuspected outcrop of rock on the road. The streets were unpaved in my youth. This calamity was termed '*bucking your toe*'. Often the nail would fly off, taking with it a large piece of skin. Then the toe would be bandaged, and look quite fetching in its white mini-turban. Such a calamity was honourable stuff, and was respected by the other boys.

But it was a different thing when a toe was infested by a chigoe (jigger) under the skin or nail. The remedy was to 'winkle out' the bloated sac of the parasite with a sterilised needle. The toe was then bandaged, and if infected it was poulticed. An infected jigger toe was a sign of neglect. The other boys could distinguish such a toe from a wounded 'buck toe', and would shout this derisive doggerel at the poor fellow:

'*Ton gimi rod, jiger de cam.*' ('Stone, for God's sake, make way; here comes jigger toe.')

This was a reference to the extreme care and anxiety with which such a handicapped boy walked lest he bucked his jigger toe! He usually walked with a heavy limp, carrying the affected foot well forward, putting his weight only on the heel, keeping his eyes fixed on the poor toe which was cocked skywards. Sometimes a piping voice (usually a smaller boy) would shriek:

'*Stone gimme road!!*' and run for his life. Or the single insulting word 'STONE!!!' would be hurled from some hidden source.

The effect was always the same—galling impotence. The victim's only consolation was the hope that one day his tormentors might be in the same plight.

As children, we seemed to have so many ways of irritating other people, by words which, while apparently innocent in themselves, nevertheless hid a deadly meaning which was known to the victim. It became a game.

Thus among the tribal peoples living in the district from time to time some man would pack on his head his worldly goods consisting of cooking pot, a piece of rag which did duty as cloak by day and blanket at night, and a rolled mat, and walk; presumably in search of other accommodation. To us this pathetic picture of a man carrying his whole world on his head was comical.

It was not long before we were shouting:

'*Fo da lili pla-ba!*'

Just those few words. But how it used to madden them!

We took care to be out of 'arrow shot'. What we were saying to them was:

'Shame on you to leave home for a trifling disagreement with your wife' (palaver; *pla-ba*).

'Silly ass! Ya, you can't take it!'

They were probably not even married.

Of those days in my native back streets I can recall only one disappointment. One evening I was out with my paternal grandmother when I saw my first white man. He was walking along Malta Street. Possibly he was sight-seeing, or perhaps he had strayed. At the sight of him I turned and stared. Innocent soulful eyes and open mouth, that was me! I was fascinated at the apparition. It was strange, and that meant only one thing—funny! Real laughing funny. I took a deep breath, opened my mouth, thought of a good one, and was just about to let my simple soul express itself uninhibitedly and forcefully, when my grandmother slapped me so hard that I nearly fell into the gutter. It was the side of my head, I think, just behind the ear.

You see, among our people *it is rude to stare*, much less to follow up the stare with naïve comments. That was my crime.

When I think of what I and thousands of my fellow countrymen have had to suffer in later life from small white boys in European cities, my only regret is that my instincts were not allowed their natural play that day. I always think of him as 'the one that got away'.

I remember, years afterwards, once asking a fellow student in London why it was that British children kept asking us the time of day. They seemed to have a strong sense of punctuality, I thought. My friend, who had been longer in the country than I, looked at me and said slowly:

'Well, it may be that, of course. Or perhaps they want to see if you will emit sounds like a human being, or just scratch your hinder parts and grunt.'

Boys will be boys.

❦ 7 ❦

Family Life

OUR day started with family prayers. Our grandmother and aunts usually joined us upstairs, and so did the foster brothers and sisters. We sat round the room, and, after the word of grace from father, each read in turn a verse from the portion allotted for the day in the Scripture Union card.

Almost unique among all the peoples of West Africa, practically all Krios are literate. The poorest family would die rather than not send its children to school even for a year, so that they could read. It is considered a disgrace among them for a child to be illiterate.

This, together with such distinguishing features as dress, language, Christian religion, English or anglicised names, the acceptance of monogamy, and even the type of their houses, all stemmed from the fact that the Krios of Sierra Leone started life as a settler community under Christian British influence.

Krio baby boys are circumcised when a week old. Baby girls have their ears pierced at the same age for ear rings. The Krios, like their main ancestors the Yorubas of Nigeria, look to the East for their basic origin. But where the Yoruba ancestors thought of Yemen, Arabia, and Mecca, the Christian Krios think of the stories of the Old Testament.

Among them, a mother, after giving birth to a child, is churched, before she can go out and resume her social activities. Each time any of my brothers or sisters was baptised, the family attended morning service at St Philip's, with mother dressed all in white. After the service the family moved up near the choir, and mother knelt at the steps of the chancel, where our pastor read the service for the Churching of Women, and purified and blessed her. Then followed the baptism of the baby.

Christianity was the central force of our home. But though a strict disciplinarian, father was not a bigot. Those foster children who were not Christians he left alone. They were not forced to attend family prayers nor to attend church. And when one after the other they asked to join us, and to go to church with us, he sent for their parents, and discussed the matter fully with them; and it was only after the latter had given their consent that he took steps to have them instructed in Christianity, and prepared for baptism. I can remember one of them, who at the time would be in his late teens, being baptised at the same ceremony as one of my baby brothers.

All this fitted in with the family prayers at home every morning and evening. On Sundays we had singing of hymns and psalms at home as well. It was a great day when I too was able to read my share of the verses of scripture with the rest of the family. Later on I became the family organist, and accompanied the singing (hymns and psalms) on the family harmonium (American organ).

But it was playtime in the afternoon and evening that meant so much to me in those early days. I looked forward to three

53

o'clock. Although I could not tell the time, I could sense when the hour approached, and would hover at the gate peering out for my foster 'cousins' to return home from school.

I can remember a long bench which stood against one wall of the house, in the yard. Round this we would sit or stand, and take part in all those pastimes which, all the world over, fill in the period between coming home from school and going to bed. Along another wall of the house were flowering shrubs.

Those afternoon play sessions in the yard must have done much to help me develop normally. As most of my brothers and sisters were not then born, but for this companionship of foster children I would for some years have been more or less a lonely child, a phenomenon which is rare in African society. As it was I grew up instead in the fold of young people.

*   *   *

Apart from this I also played with boys of my own age in our street. There was cricket, not to mention top spinning, hoop trundling ('gig running'), kite flying, and other activities.

When sent on an errand down the street I would debate whether to take my hoop or a top. If the hoop, I trundled it all the way by the simple method of hitting it with a short piece of stick in a straight line. This was the quickest method of getting to my destination.

On the other hand the top was a more interesting if leisurely and erratic way of going on an errand. I would spin the top and lash it with the cotton thong tied to the end of a short piece of stick. On its flat surface I would make marks with chalk, and these would join into lovely patterns as the top spun. Thus I would progress sideways, with my jutting posterior describing little horizontal arcs like a waddling duck, as I lashed the top on its spinning course.

What different techniques were applied to this simple pastime! At times a gentle stroke with the whip, and the top would hum forward a couple of feet or so, either in a straight line or off zigzag. There it would continue to spin. Sometimes I would let it alone until it was almost spent, and then coax it once more into animated life; then again suddenly I would go berserk, leap into the air, and give it a terrific wham with the thong, and the top would shoot up and glide in a graceful curve to land on its spinning point several feet ahead.

We often had competitions to see who could hit his top furthest in one hop. Needless to say, on such occasions the concentration was so intense that we were lost to the outside world. Once my attention was so occupied that I did not see a pedestrian coming, and I mistakenly hit his ankle, hard. He was a large man...!

Eastertime was kite-flying season, and long before then we would get our coloured tissue papers from the shops, and the special bamboo canes with which kites of different kinds were made. These varied from the simple skate design most commonly in use, to the powerful giant *Ognos* (Krio for hog's snout). This takes its name from the open triangular top segment of this hexagonal kite with its often double tail, altogether a ferocious-looking object.

Competitions were often held as to whose kite would rise the highest, or perform the most intricate loops in the sky. Sometimes this friendly rivalry would take the form of a cold war. A boy would deliberately get his *Ognos* to loop and entangle his opponent's kite. Sometimes the lead twine would snap, and we would all run excitedly to recover the errant kite, often over fields, and across several streets and properties, before the kite was caught, or fell to the ground, or more probably became entangled in the top of some trees and had to be abandoned.

As for cricket, we used the base of the street lamp post for stumps, or if we played in our yard we used an upturned pail or wicker basket (*blai*). The bat was a curved blade shaped from the giant petiole of the compound coconut leaf. As a result the commonest stroke was the hook, and the usual method of dismissal was by the catch. This was no disgrace, as usually it was the result of a hefty swipe which might send the ball soaring to the roof of the house, from which it would bounce and roll down into the patient hands of a waiting fielder!

This ball was made from one of mother's black stockings, the toe being rammed with soft cotton waste, tied hard at the neck, then turned inside out and tied tight again, the process repeated until the whole stocking had been taken up; then the edges were sewn to the surface. A well-hit ball has been known to stay lodged on the roof or in the branches of a tree. Then the game was abandoned as a draw.

Similarly, if a serious dispute arose among the players, each boy would walk off with his ball, bat, wicket, or whatever part of the gear belonged to him.

I cannot recall having many girl playmates in those very early days. I think in their early days boys tend to mix with boys. Possibly also our parents had something to do with it, for on one occasion I remember one of the little girls coming to play with us. When father came home in the evening he sent her away, and did not seem pleased. And as the eldest of my sisters was not born until I was seven years old, when I had started school, in those earliest years mine was a man's world.

*   *   *

At home, among the young people, we all called each other 'cousins'. But among the servants they were all 'brothers'.

There were the four 'hammock boys', who carried father to and from work and on his various tours of inspection of water-works installations in the peninsula. There was the 'lunch boy' who called at eleven to fetch father's hot lunch (this was called 'breakfast') and take it to the office at the Water Works headquarters on Tower Hill, and brought the remains back in the afternoon. There was another 'boy', somewhat younger, who did much of the heavy work and ran errands, when the others had gone to school.

But over and above this there was the 'head boy'. I cannot clearly recall what were his precise duties. He was a *major-domo*, a butler, foreman, or chargehand, rolled into one. He was treated with respect, and he saw to it that the others reported for duty regularly and did their jobs well.

One day he came up and said to papa:

'*Masa, a de go kontri.*'

(Please, master, I have come to say goodbye. I am going back to my country.) Papa wished him God-speed and inquired when he would be returning.

'*Waka gud! Ustem yu de cam bak?*'

But Sori, who had been with us so long that he was part of the family, had suddenly grown tired of life in the city. He wanted to go back to his people, marry and farm his land. But he was not leaving his master in the lurch; instead, he said:

'*Dis na mi broda!*' (I've brought you my brother), introducing another and somewhat younger man.

'But every time you bring a man for a job, you say he is your brother', countered my father, eyeing the newcomer carefully.

'Yes, master', Sori answered, speaking in Krio, which was the only language he knew apart from his own native Limba.

'Yes, master, we are all brothers. Same country. Same chief. But this man, he is my real brother. Same mother.' The new man took over the majordomo-ship, and when years afterwards he too left he first brought a 'brother' to take over his place.

Father abhorred smoking, and the servants would hide in all sorts of corners in the yard or in the garden to light their little short-stemmed white clay pipes. This they did by the simple method of tamping down the tobacco in the pipe bowl, tossing on top of it a piece of live coal from the fire, and making a run for it as far away as possible from the house and father's sharp nostrils. Sometimes there seemed to be in the pipe nothing but the lump of coal; but they would suck away at it, like babies at their dummy teats. Invariably if father caught them he would confiscate the pipe.

My father was six feet one and a half inches tall, and slim. Rather ascetic looking, with lean features, high forehead, strong cuboid chin, and a trimmed moustache across the whole length of his upper lip. Often, having ridden on his hammock all day, he would prefer to walk on the way home. He took deceptively long strides, stooping slightly forward, marching ahead of the men, who, even with only the lightened hammock to carry, would have to run, in order to keep up with him.

Certainly my brother Arthur and I had great difficulty in keeping up with him, years afterwards when we were young men in our late teens.

'Come on, men, don't fall asleep', he would say to us, as we marched manfully on, slightly irritated that we could not keep up with 'the old man'. Indeed there was another sore point with us, concerning our father. We never grew taller than he! However much we measured our heights Arthur and I never exceeded six feet!

*   *   *

Although ours was a heterogeneous household, yet it was in the fullest sense of the word a united family. We all loved our parents. Father's authority was ably backed by mother's loyalty. Everyone knew his duty and did it.

When difficulties arose they were settled without ill feeling or a sense of injustice. Even when things were missing, somehow the culprit was always found, or confessed.

In serious cases of theft father used a method of detection which I have not heard mentioned anywhere else. It was called the 'Bible and Key'. We all sat in a circle, and father said a short prayer. Then a small bible was opened at random, a door key was inserted into the open page, and the Book was shut and bound with some cotton or other tape in such a way that the key was firmly clasped between its closed pages leaving part of the stem and the oval handle of the key jutting out.

Then each person took turns with father to support the bound bible and key, each inserting the tip of his middle finger under the edge of the handle of the key. The bible was thus delicately balanced on the tips of the two fingers, that of father and that of the person whose turn it was.

Father then spoke the following words: 'By St Peter and St Paul and by the Holy word of God, if it is A— who took this thing (naming the lost article), let this bible turn round and fall down.' These words he would repeat three times.

The curious thing was that nothing would happen with every other member in that circle. The bible would remain delicately balanced on the two upturned fingertips, while the words were spoken solemnly, once, twice, three times. But when the culprit took his turn with father, as soon as the words 'let this bible turn round and fall down', were uttered, the bible would twist out of their fingers and drop. It was picked up again, held in position, and the words repeated; again it would twist round and drop. It was uncanny. Try as one would nothing, no willpower, could prevent this happening, when the culprit had his finger on that bible and key.

Usually the culprit would confess after this exposure. Only once did the person continue to maintain his innocence. But many years afterwards he wrote to father and confessed.

❦ 8 ❦

My Pal Jabez

'*Granny, a don cam ma.*'

(Good day, grandma.)

'*Kabo, Jabez. Wa u granny?*'

(Good day, Jabez. Is your grandma coming?)

'*E de cam ma. A lef am na mami Ancel. E se mek a tel u se e de cam ma.*'

(She'll be here soon 'ma. I left her at granny Hancel's.)

'*Ageh, yu don redi? Jabez don cam.*' (Jabez is here, Ageh! Are you ready?), the old lady called.

'*Yes 'ma*', came the reply.

'*Comot no. Mekes cam don.*'

(Then hurry up and come down!) she shouted.

Presently footsteps were heard running downstairs, and a boy of about four presented himself to his grandmother, and joined his friend Jabez. The two friends made off, hand in hand, with that sudden lighting up of the eyes which said more than words, 'Welcome, pal! I have been waiting for you!'

Together they went across to the backyard where, screened from the elders, Ageh took out of his pocket, with difficulty, a large ripe mango twice the size of a peach, yellow and golden and red in colour, and succulent. His friend's eyes brightened at sight of the luscious fruit.

'*Usi yu get am?*'

(Where did you get it?) he asked.

'*A pik am.*'

(I picked it.)

'*Usi?*'

(Where?)

For answer the other took him to the fence separating the yard from the property next door. Over this hung several branches of a lofty mango tree which grew next door, sixty feet high.

'*Luk am*', cried Ageh, with pride, pointing to where the fruit had hung. There were other specimens not quite ripe, still hanging over the fence into their yard.

The branch was only about ten feet high; but to a boy who was himself less than four feet tall it was a magnificent achievement to have downed that fruit; and the look of admiration tinged with envy in his friend's eyes showed that the achievement had been duly appreciated. Hiawatha in his first hunt could not have felt more proud.

It was one of the advantages of living in Kossoh Town that mango trees were common in the gardens there. Boys living in the central districts of the city were not so fortunate. These lofty trees and their golden fruit had a magnetic attraction

for boys of all ages. The first target in a Kossoh Town boy's
life was to down his first mango.

This was a feat which came only with practice, usually
under coaching by an older boy. It called for steady eye and
balanced muscles. The stance was all-important. You took
your position a little distance away, your left foot pointing
towards the target and the weight of your body carried lightly
but firmly on your right leg. The latter must be springy, bent
at the knee. With the stone held in your right hand, you
drew the latter as far back as possible, almost touching the
ground, and then with a whip-lash recoil of the whole body
you swung the right arm and hand over your head, bringing
the whole body into the motion with a follow-through which
sent the stone curving in a giant parabola, slicing the stalk
of the fruit as it neared the top of its trajectory. Then, its
mission accomplished, the stone fell to the ground.

It took practice, a good eye and steady aim to become adept,
and it was quite common to see a battery of children attacking
a tree, often for just one fruit dangling temptingly. Rivalry
was common. But sometimes a parent or an irate owner
would intervene. This hazard made it a real adventure. So
Ageh was lucky in having in his back garden his own branch
on which to practise his first attempts.

He took a big bite and passed the fruit on to his pal, who
took an equally big bite; and soon they were lost to the world
until they had gleaned the last vestige of fruit from the large
flat 'stone' core. Then they returned to join their grandmas.
Jabez's grandmother had by this time arrived, and was deep
in conversation with Ageh's. The boys stood silently by their
grandmothers, awaiting their pleasure.

Jabez's granny took one look at them and without the least
emotion she rapped her grandson sharply on the head.

Immediately the lads knew that something was wrong and they retreated into the backyard. There it suddenly dawned on them that they had forgotten to wash their hands and their mouths after their feast. So they went to the pump in the yard and did a quick toilet.

Then they returned and took up their position beside their grandparents again. After talking for a little while the latter turned round, looked at their grandchildren again, and this time both women rapped their respective grandsons. Harder.

The lads disappeared again to the backyard in consternation.

'What is the matter now?' they asked each other.

And then Ageh noticed that Jabez's coat had some stains from the juice of the fruit. As he pointed agitatedly his friend was doing the same to him. Both of them had stained their coats with the yellow juice of the mango! This was a terrible tragedy; and they went and got a towel and soaked it and scrubbed at each other until they managed to get the stains off. They were wet to the skin. But they couldn't help it; and in any case the sun was glorious, and it would soon dry.

Rather worried they returned to their grandmothers, who had now donned their hats. Each lady caught hold of her grandson and, making sure he wouldn't run off, gave him a sharp final cuff, as if to say: 'You mind your step next time, silly boy!', and off they went into the street. There they let go of the children and the two elders walked ahead, while the boys romped behind them.

They went a few yards up Pownall Street and turned right along Malta Street at the corner where stood Mrs Hancel's house and small shop. The ladies exchanged the time of day with Mrs Hancel and continued on their way, past the corner where Mammy Cromanty's house stood on the left hand,

and exactly opposite on the right hand that of her famous nephew, Abamba. This gentleman was an object of awe to the two young friends, for, although a professing Christian Krio, he was known to be a member of the dreaded pagan Yoruba *agugu* cult—those devils clothed from head to foot in red gowns like mahogany Ku Klux Klan, with switches of horses' tail in their hands, and whose eyes were said to ferret out little boys from whatever corners they might be hiding in. Could this be the same man whom they knew as 'Uncle' Abamba, with whom they exchanged 'good days!'? Could he really turn himself into this devil? They hurried past his house, furtively looking into the open door, expecting to see they knew not what.

At this corner stood one of the public standpipes of pure water provided by the Water Works Department. One stood in almost every street. But at the moment it was dry. For this was the height of the dry season, the hottest time of the year, when the paps of Mother Nature dried up with the intense heat.

It was a glorious day. The sun was intense, and hardly a wisp of cloud trailed aloft, while under foot the ground was scorching to the bare feet of the two boys. But the little group did not notice these things as they trotted along past the '*ajoini*' (rows of single-storied two-roomed flats) where lived the Campbell family: a mother, her two daughters and three sons Festus, Tommy and Dick. All three sons were stalwart members of St Philip's choir.

The street crossed the stream which, during the rainy season, cascaded here from the slopes of Mount Aureol, but was now barely a trickle. Immediately past this bridge they came on the left to the filled-in public well which used to serve this part of Kossoh Town. On the right, exactly ten feet opposite the well, was the local public dustbin, a small square shed with corrugated metal roof, and the top half of its sides open. It was filled with all the rubbish of Kossoh Town, rags, remnants of food, bones, worn-out baskets, and the carcass of a dead dog on which a couple of *Yubas* (vultures) were feasting. One very much alive cat foraged intently in the contents. It was a hot day....

Presently they reached Patton Street, turned right, and entered the gates of St Philip's, Patton Street, Parish Church of Kossoh Town, passing underneath a brilliant canopy of bougainvillea and trees bearing large bell-shaped flowers of delicate ivory cream and palest primrose hue. They crossed the churchyard and entered the Sunday School building.

Here was assembled a number of ladies of all ages, but practically all married, and either matrons or grandmothers. It was the weekly Dorcas meeting. Half an hour later a strange sight could be seen. The roomful of august women sat sewing assiduously with just the minimum gossip necessary to keep things moving smoothly.

And in the midst of that throng sat two little boys, each barely four years old, sewing, ever so seriously, a doll-sized *cabbaslot* (Krio frock), which each had cut to pattern, with a pair of scissors which their grandmothers had given them.

That was how Jabez and I attended our first Dorcas meeting.

9

Kossoh Town Folk

MOST of the ladies at the Dorcas meeting seemed to be about the same age as Jabez's grandmother and mine. It was a grand sewing meeting of the elder women of the parish in the tradition of the early Christians who used to meet in the house of the saintly Dorcas. The things they sewed were sent to the Sierra Leone Missions for use in the hinterland of our country.

Our church was referred to affectionately among us as Granny Church. It was later that I knew it by its official title of St Philip's. There were many grannies in Kossoh Town. In addition to our own two, there were at least eight grannies in the quarter-mile stretch of Malta Street leading to the church. There was granny Cromanty, Esi granny, Duro granny, Mammy Bedford, Mammy Moses, Mammy Bultman, Mammy Metzger, Joko mammy. We knew all of them, and addressed them as 'granny' or 'mammy'. Their daughters, who were of the same age as our own mothers, we addressed as 'aunty', 'sister' or 'sissy'. Their grandchildren were our contemporaries and playmates.

The adult members of these families were almost all females. There were very few men among them. Now and again some 'uncle' would arrive from 'down the Coast', as the other territories of West Africa to the south of us were called. From time to time our playmates had baby brothers and sisters. But we never saw their daddies.

As for grandads I do not remember any. We ourselves had an Uncle Jacob (father's uncle), but our grandfathers on both my parents' sides died long before we were born. The same thing had happened to Jabez's grandfathers.

In later years it became a good-natured rivalry between the surviving grannies as to who would live the longest.

'I see you are still going strong', granny *A* would say to granny *B*.

'Yes, by God's power I shall live to a hundred yet', granny *B* would return the banter.

In the end our granny Cole lived to the ripest old age of all, ninety-three; but the last to pass away of that noble band was granny Cromanty.

*   *   *

At about this time I would be between four and six years old. I did not start school until two months before my seventh birthday. Unlike my earlier infancy I distinctly remember my companions of this period, even though separated from many of them by distance and time.

Jabez was the first and most intimate of this group, and I recall this period mainly by the adventures which we two shared.

Those were wonderful days, that in-between period when I was beginning to understand what was going on, but was not yet ready for school. Somehow my brother Arthur does not seem to have shared this period of budding outdoor activity. I suppose because he was two years younger than I, and when you are five or thereabouts, three is very young! Jabez and I were like twins. There was only two months between our ages.

We played, fought mock battles, wrestled, turned somersaults (*tonobo*). Sometimes, when his grandmother was away, he would have his meal with us. On the other hand theirs was the only house where I was allowed to accept any food or present. We compared ourselves, our heights, the speed with which we could run, how high we could climb, and even our bodies. As is natural among boys at that age we compared notes with other boys of similar ages and soon found that some of the boys of Kossoh Town were like ourselves, while the others were different, although we did not then know that this difference had to do with what the grown-ups called circumcision.

Actually from the very moment we started school our paths diverged; we never attended the same school, nor did he go to college as I later did. But he was my first playmate, and we remained lifelong friends.

This was perhaps because our church activities bound us closely together. We attended church together, and the same boys' class on Monday evenings. We accompanied our grandmothers to Dorcas meetings, joined the choir about the same time, moved up to the Tuesday evening class, and were confirmed together. If he was sent on an errand which took him within a quarter of a mile of our home, he would manage to call, and if I on the other hand were in his vicinity I would call at his place.

The reason for this close friendship was probably because our two families were very friendly. Apart from our grandmothers, our two fathers were also close friends. The two grandmothers were tall and slim. But Jabez's father was of medium height and somewhat burly.

He was a master printer, and among other things printed the hymn sheets and concert programmes for our church. I loved to accompany Jabez to his father's shop. How fascinated I was by the pockets full of the small leaden types! How we would watch as his hands roamed deftly, selecting the pieces, and setting up the type! Then he would transfer the forme to the hand press, feed the paper, and work the large wheel which printed the sheets. He would often give us the first two copies he struck.

This printing shop was at Fourah Bay Road, a little beyond its junction with Bombay Street. This was a busy commercial area, where a number of well-known citizens had their shops.

This busy area always attracted us children, especially on a Saturday evening, when the kerosene lamps of the women sellers mingled with the incandescent lights of the shops to produce a thrilling sight.

Around the inevitable standpipe of fresh water at this junction would gather the throng of women and children fetching water for their families. Near here also was the provision shop (grocery store) of Mr Amadu Taylor, barrister and businessman, whose uncle incidentally was the father of Samuel Coleridge-Taylor the famous Negro musician. He had a noted barrister son and was the first African to send his daughter to the United Kingdom to study medicine. Unfortunately she died on the eve of completing her training. This was before our time. But his shop was a mecca for us children. A penny there was well spent.

*　　*　　*

Now that I was able to venture away from home, our own Pownall Street was beginning to have a meaning for me. And we were a very mixed community of all tribes and customs. At the top of the street lived Mammy Hancel. From time to time she was joined by her grand-daughter, who came from somewhere 'down the coast'. She was very haughty and treated us boys like pariahs.

Next door to them, and exactly opposite us, was a house full of native people of Timne and Limba stock, presided over by one Mammy Sama. This was a mud and thatch house whose forecourt was continuous with the street. Here on certain nights the men would dance the conga till after we had gone to sleep. The people of this house were scantily dressed. The children of both sexes went naked, the women had their bosoms bare.

Here also for many years a man called Pa Sori used to sit in a chair in the sun, and as we went past we would greet him, as we did everyone else in the street. As the years passed his hands and his feet became more and more deformed until gradually they were merely red stumps. His fingertips and toes went one by one. His nose was gradually eaten away. It was not until many years later that I knew that this was a case of leprosy with which we had shared the same portion of the street all those years.

The next house was Duro granny's and her daughter and grand-children. They were Krios like Mrs Hancel and us. Next was a vacant plot, which was usually planted with kassava and Indian corn, and next to this was a big compound belonging to an important Mohammedan Timne headman, an Alimami.

Unlike the people in the house of Mammy Sama and Pa Sori, these Moslem Timnes were dressed in the full robes of people of this faith. Every evening they would assemble in the yard, unfurl prayer mats, and say their prayers. Now and again the headman himself would be seen in the street on some errand, always with two or three of his wives walking behind him. They were well dressed in *lapa*, *oja*, and *buba*, with beautiful sandals, and gold rings, bangles and ear rings. At the feast of Ramadan the prayer activities were intensified, with feasting at sunset, and festivities at the end of the period.

The rest of our street contained this same mixture of

Krios, Timnes, Limbas, Mendes, Moslems, Christians, pagans, poor, well-to-do, traders, shopkeepers, and people without visible means of support.

Thus from earliest childhood I was used to all the various elements of humanity which made up our country. We grew up together, played together. Even though afterwards our paths diverged as we grew up, yet nevertheless they have remained a part of my life.

This advantage of a cosmopolitan childhood was a peculiarity of boys who lived in the east end of Freetown, as distinct from those in the west end and central parts of the city. It goes back to the history of the founding of Freetown. As the settlement flourished, and indigenous natives moved into it from the hinterland, the wave of Protectorate immigrants first hit the east end, especially Kossoh Town. It was only later that it spread over the rest of the city.

*　　*　　*

My father lived among them all, friendly with everybody, respected by all, and yet maintaining his family intact in a way which could not have been more complete had we lived in an isolated area, or 'among our own class'. In a way our life was all the richer for the diversity of our neighbourhood.

Now and again someone would call on father dressed in flowing robes, obviously an important member of the local Muslim or native community. This was usually in the morning. He would take a seat in the parlour, and father would have a word or two with him, then go and continue with his preparations for work, stopping from time to time to have a little more chat. Not a lot was said, but apparently these meetings were greatly prized on both sides, and promoted mutual understanding and friendship.

At other times some richly dressed women, usually in pairs, would call bringing gifts. I remember how very beautiful they were. They were the wives of a Fulani chief who lived in Fourah Bay Road opposite Dovecot. Apparently they were sent by their husband and lord to present his compliments and good wishes to father. Now and again on his way home from work father in turn would call on the chief to return the compliment.

Father, a strict Church of England Christian and an import-ant citizen, was on the friendliest terms with everybody, from the most orthodox Muslims to the most nondescript pagan native. He was the peacemaker, arbiter, and indeed the unofficial but much respected leader of this heterogeneous community of Kossoh Town.

A somewhat austere man, his popularity was immense, and he was affectionately known as 'Water Works Cole' through-out Freetown. As for the grannies of Kossoh Town they called him, simply, their Wilfred.

There was a reason for the 'Water Works Cole'. Although the Freetown Water Works is the oldest on the West Coast, yet it runs dry every year during the three months of March, April and May, the driest and hottest season in Sierra Leone. When that happens the water is turned off at source, except for a few hours at dawn and dusk. The result was that although there were public standpipes in every street, the queues were fearfully long.

At such times father always opened the gates of our home and allowed the local people to come in and fill up their pans and buckets at the standpipe in our yard. We paid water rate according to the number of taps we had. The long lines of altercating women and children would waken us in the early blush of dawn. Father was the Superintendent of the Water

73

Works, a municipal undertaking. But they did not blame him. On the contrary they seemed to think that if with him in charge we still had a drought, then nothing could be done about it.

We often had visitors to our house, usually some crony of our grandmother's, or some 'cousin' or 'aunt', who lived in another part of Freetown, or in one of the outlying villages. Waterloo, Benguema, and Hastings were names of villages which cropped up often in this connection. African ties of kinship are both elastic and extensive.

Invariably grandmother would offer them refreshment. This usually consisted of *agidi*, and she would send me to fetch one. I would go to the larder and fetch one of these small bundles wrapped in leaves and looking like a miniature cocked hat. She would unwrap it, revealing a solid white blancmange-like substance made from the finely ground and sifted flour of corn, boiled into a gel, and cooled into shape. This she would turn out into a basin, and with a large spoon would crush into a mash, adding sugar and water from an earthen cooler. The result was a delicious cool drink, very refreshing to her guest.

❧ 10 ❧

Preparing for School

AND so the time came when I was due to start school. Long before then I had been impatiently longing for the day. At home Shenkoo and the other foster brothers and sisters had all been going to school. Each morning a procession of lucky people left for the outside world of romance.

First the older boys and girls went off to Bethel, the school attached to the Methodist church of that name, and Kossoh Town's own school. Then father left on his hammock or palanquin with his four boys. Sometime during the morning the chop boy came to fetch his lunch. The house boy went to market and returned. In the afternoon the boy with the lunch box came back, then the boys and girls from school, and eventually father at varying times, depending on whether he had been out of town on his duties, or had been attending one of the very many committees on which he served.

Jabez too had started school at Bethel; and so had a number of other boys and girls of my own age. They were now in an altogether new world, and I too longed to enter it.

Not that school was an unmixed blessing. From the tales which Jabez and others brought home, I gathered that it was a place where you were likely to be caned mercilessly almost every day. The headmaster at this time was a man we all knew, and one whom I could not possibly believe to be an ogre. Tall, slim, and mild-mannered, he always returned our greeting when we wished him 'how do'. Except that he generally seemed absent-minded, we felt that he would like to pat our heads as he passed. Certainly we were not frightened of him out of school.

But this was the same man who, Jabez and others told us, became a living devil once he was at school. The school was just one main hall divided in two by a wooden partition. The classes were open to each other, so many benches taken up for each class. The only dividing mark was the teachers in front of each class. So everybody saw what was happening in all the other classes, including canings which were very frequent and, according to my pals, violent.

I paled at what might lie in store at school. But the attraction for me was too strong.

On more than one occasion I asked father if I could start school. But he said:

'Wait.'

The more I waited the more I wanted to be like the other boys. This was the harder to bear as Bethel School was in the very next street to ours, and not more than 500 yards away. At lunch time and at the close of the day, I could hear the sound of laughter of the happy band of children at play there. I longed to join them. Not getting any results from father, I besought my mother to intercede with father. But she too said, 'Wait a little while'.

Then my father, who always explained things to us, called me one day and said:

'Now, Ageh, why do you want to go to school?' This was an easy one to answer.

'Because Jabez and Jonah and Taiwo and Moses are all going to school, sir.' These were boys of more or less my own age. 'And Shenkoo and Santigi too', I said, adding the foster brothers from our own home, who, however, were older than I.

'I mean what do you want to do at school?' my father insisted.

'I want to learn to read and write. I want to carry my slate.'

The carrying of a writing slate was the hallmark of boys and girls attending school.

'Yes, but when you finish at Bethel School what are you going to do?' he asked. 'You see,' he continued, 'once you start school you will have to go on and on for a long time until you are a grown man.'

I opened my eyes wide. I was too young to understand that education was a ladder reaching far beyond the Bethel School stage. All I wanted was to be like the other boys.

'It is important which school you start with, because when you have finished there you have to go to the Grammar School', papa continued.

The Grammar School! This was different. That was the most famous school in Freetown, where father himself had gone.

'We must lay a good foundation', he said, adding, 'when you start school you must study hard, and if you do, who knows but you may go to England one day!'

England! I was really interested now. For even in those my early days, England was a magic word to us. It was the country where our bishop, our governor, the women missionaries, and the white men in the shops all came from, and where our own people had to go to become doctors and lawyers. It was almost as far as heaven in our imagination, except that heaven was above the clouds.

How right father was about Bethel School I was to realise in later years, for I cannot ever recall anyone going on from that school to a secondary school and college or a profession. What it did for Kossoh Town was to provide the means of elementary literacy for the local people. It was near at hand, and fees were low.

Then one day father said to me:

'Ageh, would you like to start school after Christmas?'

Would I indeed! Those few magic words made that Christmas stand out in my memory, even after all these years. The preliminaries for starting school, Christmas preparations in the church, Christmas Day festivities, and the whole Christmas holiday, are a patch of green in my garden of memories.

77

The first thing we had to do was to visit Uncle Willie. Uncle Willie was a half-brother of father's, and a master tailor. His shop was in the central ward of the city at the bottom of Garrison Street, between Kissy Street and East Street, where the railway line passed on its way to the interior of the country. It was the first time I had been to this shop, and it was about the first time I had walked so far from home. Uncle Willie made all our suits, but usually he called at our home to measure us and take instructions. But this day it was a special treat for me, as, together with Arthur, I accompanied my mother the two miles to the shop.

Uncle rose from his machine at the open window and met us at the open door. A tall man, even slightly taller than father, and somewhat older, he wore an open waistcoat, on which were stuck many pins, and he carried a tape measure draped over his neck like a minister's stole. He greeted us warmly and offered us seats, but I was too interested in the busy street outside, and preferred to stand at the doorway looking out. Just below his shop was the busy covered market where so many people were going and coming, with purchases of meat and other foodstuffs.

'So you are going to school, Ageh?' Uncle Willie commented, as he proceeded to measure me for my suit. When he had finished he showed us round his shop. It was of two rooms. The front one had three treadle sewing machines, one in front of each window, presided over by uncle and his 'journeyman' (qualified assistant). A third stood further back and here sat an apprentice sewing. There was a second apprentice doing some work in the other room.

Occupying almost the whole width of the shop was a long cutting table piled with lengths of cloth, including a piece of material which uncle was marking with chalk and a long wooden measuring ruler. The second room at the back of the shop was store, retiring room and rest room combined. Uncle gave me biscuits and poured some water from the cooler. It tasted cool and refreshing after the hot walk. This cooler, which was like ours at home, was in the shape of a huge cockerel, grey in colour and of unglazed earthenware. Its surface was damp and sweating, and felt cold to the back of my hand. Lifting it by the handle, uncle poured the water through the spout of the cockerel's beak into an enamel cup bearing coloured portraits of the king and queen (they must have been King Edward and Queen Alexandra) with crossed Union Jacks between the two. These cups and mugs were common in Sierra Leone households for as long as I could remember.

From this start, the visit to Uncle Willie, the excitement of that Christmas grew, as the weeks passed by. 'I am going to school', I kept saying to my foster brothers and sisters, and, somehow, I felt that from that moment I was really accepted as one of themselves. I was promoted to that magic band of school-goers. I counted the days until I should march out with them one morning to that outside world, which beckoned beyond the gates.

And then came the day for scrubbing out the church. Jabez and I accompanied our grandmothers to this function as in previous years.

The church bells started pealing early in the morning and were repeated at intervals. Soon we set out, grandmother and I and Arthur, carrying buckets, soap and scrubbing brush, and a small enamel basin for scooping up water. At the church we met a concourse of other women of all ages, and children of both sexes. Soon the task began.

First of all, aided by some volunteers of the men's class who had come early before going to work, the broad loom of coconut matting was rolled up from each of the three aisles, and carried to the churchyard. Then we lifted some of the pews into the churchyard and the street. These pews were rows of chairs battened together into forms, which were slotted into wooden rails nailed to the floor at the side of the aisles.

The whole church was then swept, each of us taking a hand. We were now ready to start scrubbing.

The young people fetched water from the pump which stood just outside the church, and first the pews were all scrubbed completely with scrubbing brush, soap, sand and water. The backs, the tops, the seats, and the legs were all scrubbed. Then we turned to scrubbing the floor of the church, and washing down the walls. The floorboards glistened white with the fine grains of sand, and the whole building smelt freshly clean. The windows were thrown wide open, and by late afternoon the church and vestry were clean and dry from the heat and the sunshine.

Then came the procession from the seashore. When the scrubbing was done all the young people went down to the seashore, which was only a few hundred yards down Patton Street, across Fourah Bay Road and the railway line. There, filling our pans and buckets with fresh clean sand, we filed back in procession up the shore, singing as we approached the church. One of the songs we sang I can recall:

> Home again, home again!
> When shall I see my home?
> When shall I see my native land?
> I'll never forget my home.

Not very African, but how we enjoyed it! Singing lustily thus we arrived at the church. There we poured the sand on the ground at the foot of the steps leading to the church, the vestry and Sunday School building. The yard thus had a fresh carpet of glistening yellow sand. Throughout the day our pastor paid several visits to cheer us at our work, and towards the end the wardens and other members of the Church Council came to thank the women, and help in the final heavy work.

On Christmas Eve it was the turn of the men. This time Jabez and I accompanied our fathers. It was the men's job to decorate the church, with palms, flowers and green leaves, bunting and banners. This involved climbing on ladders. Christmas Day saw the whole parish keyed up to the celebration of the Birthday of our Lord. I shall never forget that day.

Arthur and I rose up bright and early. It is sunrise by five, but we were up even earlier that day. The first thing we did was to go and open the parcel of suits which Uncle Willie had sewed for us; a simple coat and knee-length trousers. We cleaned our black boots. Then we went to have our bath, scrubbing each other from head to foot with *sapo* and black soap, under the shower. We got dressed, and were ready for family prayers which included a foretaste of the hymns which we would be singing in church that day. After this we had our breakfast, and were ready for church.

At a quarter past eight the bells of St Philip's began to ring a happy soprano, which could be distinguished from the other churches in the vicinity, the bass of Bethel, the mezzo-soprano of Holy Trinity, and the alto of All Saints.

As we neared the church, the neighbouring streets were filled with approaching worshippers, men mostly in black, women in white, young people, children, babies. Those women who went barefooted during the week now wore carpet slippers, hand knitted in brightly coloured wools. Those who wore carpet slippers during the week now wore shoes. During the week most of them wore head ties; now many of them wore white hats, either on top of head ties or directly on their heads. Of the boys, a few wore boots, the rest had washed and scrubbed their feet shining brown. Almost everybody wore a new suit or a new dress. Christmas was the occasion when new dresses were 'baptised'. Soon the church was full, some people standing outside.

Almost abruptly the bells ceased, and we all turned round to the west door, where the procession of choristers had appeared. In ringing tones pastor announced the Processional hymn:

> Christians, awake, salute the happy morn,
> Whereon the Saviour of the world was born!

and the whole church rose as one, as the organ pealed out the notes. Then the procession of choristers moved up the centre aisle, leading the singing. First the little boys, then the older trebles, and altos, tenors, and basses, all men; then followed the lay readers, the curate and pastor and last of all the wardens.

I have never heard anything to compare with the singing at St Philip's. Organ choir and congregation blended in cne mighty chorus, in which somehow the inspiration of the individual worshippers could be sensed.

I was more than ever determined that one day I too should be a member of the choir.

As the service proceeded I fell into a reverie. I looked forward to what was to come afterwards, the feasting, the *bon-bons* and sweet biscuits of different shapes and kinds, specially imported from England for Christmas in large square sealed tins, the ginger beer, gingerbread, rice bread, ham, specially home-baked cakes, and above all the Jollof rice, of pork, chicken, beef, tomatoes, oil and seasoning, the richest dish of our people. I thought of Jabez and our other friends who would be coming later to spend the day with us, the presents, the visit to other relatives to wish them a happy Christmas, and to receive little presents of fruit or sweets. I must have fallen asleep, for the next thing I knew I was dreaming I was being bitten by ants. I woke with a start to find my grandmother pinching me to keep me awake.

'Stand up!' she whispered.

She must have been trying to keep me awake for some time. I stood up, in disgrace, but perked up when I noticed two other boys in other parts of the church also standing up.

I looked about cautiously. A sea of upturned black faces was raised in earnest concentration in the direction of the pulpit. Most of the men and women had fans in their hands, which they waved slowly and rhythmically to cool the perspiration which streamed down their faces and necks.

At the pulpit stood our pastor preaching earnestly to us. He told us about the baby Jesus. I knew what he meant; we too had babies. He talked about heaven. He told us children to be good, and to obey our parents. Again I knew what he meant. And as he talked I felt that I too, when I grew up, would like to be a minister. It was the greatest thing I could think of. And soon I should be taking the first step towards my goal. I should be starting school any time now.

Bereavement

BUT it is when we are most happy that something often goes wrong. And it was about this time that our little family suffered its first bereavement, a rare event, thank God.

Although I was not yet at school, I was now being taught at home by our mother to read, write, spell, and do simple sums. Two snippets have stuck in my mind from those early days. One was: 'Bob is a big dog.' A picture of Bob the dog accompanied the story. This stuck because my own name was Bob, and our dog Fido was not only big, but also happened to be older than I. Whenever I was inclined to get out of hand I would be cut down to size, as the Americans say, by a reminder that even Fido was older than I.

The next couplet in my reader which I remember was that about a certain boy who was anxious to reach his sixth birthday, and was so proud when that event arrived; as what child would not be!

'Oh try me father, try me; I'm six years old today', went the words. There was also a picture of this boy, measuring himself back to back with his elder brother, to compare heights. I was myself six at the time....

The readers were in English, and by this time I was reading not only the first but the second and third grade primers. Some of these dealt with nature subjects. One was particularly interesting, and among other things contained a chapter on bees in words of one and two syllables, complete with pictures of the three types, queen bee, drone, and worker. It so happened that just then we had an invasion of bees, which swarmed and settled under the eaves of the house, and became a nuisance. So one weekend father got the servants to smoke them out.

My brother and I thought this was a good opportunity to learn about bees at first hand. So we went among the dead and dying bees and picked one up that looked dead enough, took it upstairs, and with my book open at the page of illustrated bees, placed the bee by the side of the pictures. We were deep in the business of identifying the different parts and finding out whether it was a queen bee, a drone or a worker, when a terrific pain seared through my fingers...it was a shocking sting!

I was inconsolable all that day.

Soon there was a third brother to join Arthur and me. Two years separated him from Arthur just as the latter was two years younger than I. He was called Wilfred, after father. I had read stories about storks bringing babies, and had seen pictures of them. But we never fell for that, for it was not one of the things told us by our own grown-ups. It was not part of our African folk-lore or tradition.

We had not known a baby was coming, but one day suddenly everybody was busy in the house. Both my grandmothers were there, and other elderly women came and went. And for once we were barred from our parents' rooms.

Then towards evening there was a strange cry. A baby's cry. We rushed upstairs, tiptoed to mother's door and knocked. But we were still kept out. Eventually when we were allowed in, there, lying beside mother, was a brand new baby, yellow-brown like a newly minted penny, its eyes not quite shut, as if it was peering suspiciously at us.

Arthur and I crept up to the bed, looked at mother and the little baby. Then, shyly, we stretched out and touched baby's mouth, his little squashy nose, and his forehead. He was alive. He was one of us.

Baby Wilfred soon proved that he was very much one of us, the most popular member of the family. He was always smiling, always in the midst of things. Soon, wherever Arthur and I went, he went; always doing what we did. We were a happy trio, three musketeers of mischief. We had measles together.

As a precaution against damage to our eyes from the measles, we had large circles painted round our eyes in some chalky material, and as we looked at each other we had great fun. The nature-study book came in handy once more with a picture of an owl. We looked like three owls in a nest, and kept shouting 'owl' at each other.

The three of us would huddle in dark corners, or under father's bed which was higher than mother's, and listen to the rain. It was cosy to hear the heavy tropical downpour beating fortissimo on the corrugated galvanised iron roof, the space between roof and panelled ceiling providing a magnificent sounding board. And when, as not infrequently happened, real tropical thunder and lightning struck, we little children actually loved it!

When Arthur learnt the alphabet, baby Wilfred joined in. Father had a large illustrated alphabet printed on a sheet of cardboard, capitals on one side, small letters on the other. It was not long before Wilfred could read all the letters from A to Z. What was more, he could pick out and identify the individual letters. And he was scarcely two.

I know he was only two, because, not so long after this, tragedy struck our home. One day there was a hush. Big tears appeared on mother's face; and father did not leave for the office. Relatives appeared. And then it came out: Wilfred was dead.

He had been well the night before; but now he was gone. A sudden attack of fits was the cause.

The next year, three years after Wilfred was born, and seven years after me, a baby girl came to our home, the first of our sisters. She was named Phoebe Winifred. The 'Phoebe' was after our paternal grandmother.

❧ 12 ❧

First Year at School

I CANNOT remember the details of my first day at school, except that I seemed to be the smallest boy in class. Probably in age, certainly in size. There were about six girls, out of a class of about thirty. I was placed in Class I of the Junior School, and so missed the Infants Department, which I could see across the corridor, through two doors which opened out from our class.

This Infants Department was under the charge of a European lady, a Mrs Mavrogadarto, who was the wife of the Commissioner of Police. The Junior School (primary department) was also under the principalship of a European, a Welshman named Holloway. Both principals were administrators. All the teaching was done by our own people. Come to think of it, throughout my school career, I was never taught by a European, but always by African teachers. Those in the primary department of this, my first school, were all men. Mr Erasmus Cole, Mr W. E. D. Campbell, Mr (later the Rev.) Lucas, Mr Doherty, and the Rev. W. T. Thomas.

In the Infants Department the teachers were similarly all Krió ladies, including Miss Amanda Richards and Miss Phoebe Joah. The building was a single-storied structure in the form of the letter 'H', and stood in spacious grounds. One limb of the 'H' housed the Infants Department, the centre piece housed two classes of the Junior School and the other limb contained the remaining three of its five classes.

At first I used to go to school in the company of Mr Erasmus Cole. The school was almost three miles away, and the only transport was our own legs. Mr Cole, who later became the Rev. E. W. B. Cole, M.B.E., M.A., and throughout his life has been among the leading educationalists and Methodists in Sierra Leone, was then starting out on his career, as a young assistant master, fresh from Fourah Bay College, where he had obtained the B.A. degree of Durham University. This post at the Government Model School was his first assignment. He lived with his family in a fine house in Malta Street, a few hundred yards from us.

Fourah Bay College, founded in Freetown in 1815 as a training institution for Africans by the Church Missionary Society, had been an affiliated college of Durham University since 1876, and most present-day leaders of West Africa have passed through its portals.

I used to call at Mr Cole's, and walk proudly and confidently by his side, my little hand in his strong grasp. He had, and has, a rasping, or what is termed 'gravelly', voice, somewhat like Louis Armstrong of Satchmo fame. It was a long walk to the Model School. No buses; no hammocks for us. No cars yet, no taxis, not even bicycles.

Freetown bears striking evidence of its settlement origin in the arrangement of its streets. They run parallel and straight. Starting from near the seashore they go straight up

88

to the foothills of Mount Aureol. To get from one part of the city to another you walked up one street, then hopped across to the next parallel, rather like a fly crawling up a wall, a crab making for its favourite cranny, or a rock-climber stretched out on the face of a precipice.

From Mr Cole's house we went along Malta Street, crossing in turn Easton Street, Pownall Street, Crook Street and Mercer Street, until we came to Patton Street. Patton Street was a north–south parallel. We followed it a varying distance, then hopped to the next parallel, Bombay Street, and on to the next, Mountain Cut.

This last was our main meridian of navigation. It was over a mile long, and by keeping to it we came inevitably to the lower reaches of Mount Aureol, and Bishop Elwin's Memorial Church, standing requiem guard over the vast Circular Road Cemetery, the original *White Man's Graveyard* of history and fiction. By the time we had circumnavigated less than two sides of this cemetery, we were practically at the gates of the Government Model school.

The return journey followed in reverse. This was not always taken in the company of Mr Cole, as a number of boys came home in my general direction, especially those living in Foulah Town and Fourah Bay, both one hundred per cent Moslem districts. Thus many of my earliest school friends were Mohammedans.

Once arrived at school, I joined the other boys waiting for assembly. We were marshalled in the boys' quadrangle by class, and then marched smartly in:

'Left, right; left, right!...left!...left...left!!...LEFT!!'

We were most of us dressed in khaki; not that it was a uniform, but it was hard-wearing, and did not show the dirt so readily. The Moslem boys wore their gowns. We

89

were a fine sight. Then followed prayers for those of us who were Christians, with singing. At the end of the school day we closed similarly with hymns and prayers.

Many of those hymns have remained nostalgic through the years:

> Forth in thy name, O Lord, I go,
> My daily labour to pursue;
> Thee, only thee, resolved to know,
> In all I think, or speak, or do....

I liked this hymn—permit me to say I loved it, for the sense of uplift it conveyed, its wide sweep of musical phrasing, its happiness and confidence of tone. It suited the mood of young people singing in the morning at the foot of one of the most romantic mountains of Africa.

Then in the evening this hymn:

> Now the day is over,
> Night is drawing nigh,
> Shadows of the evening
> Steal across the sky....

Simple, sweet, cadential, it reflected the spirits of children who were preparing happily to return to the love of mother, home, brothers and sisters.

So also were the songs we learnt at singing lessons, presided over by Mr W. E. D. ('Sonny') Campbell. A great, wonderful, and kind teacher, who incidentally was partially deaf! Of these songs none has retained for me such lingering tenuosity of sentimental pathos as:

> White sand, and grey sand,
> Who will buy my grey sand?....

I realise that I am talking like an adult now. That is inevitable. We feel best as youngsters. But, alas, we need the intellect of maturity to analyse what we felt. We depend on memory. I am glad that mine seems to have been happy.

All these songs were either English or British, including that one about Britannia ruling the waves:

Britons never, never, N-E-V-E-R shall be slaves!!

We did not feel strange about these songs. We were British, if not Britons. They were part of the treasures of the language in which we were being educated. And slavery had meant a lot to our race.

★   ★   ★

Lessons followed each other in quiet regularity. Boys continued to be boys, girls were girls, and we were all rather naughty, but altogether a typical average group of children.

Certainly our teachers looked anything but harassed. They arrived in the morning with a jaunty air, took out their canes, laid them ostentatiously on the tops of their desks, and proceeded to teach. They talked to us, wrote and drew on the blackboards, and, now and again, called on one or other of us to answer questions or make comments.

We had one short break in the morning and one longer midday recess. On these occasions we spilled out into the grounds.

The whole compound was grassed, and accommodated us easily. In the centre of the campus stood a spreading cashew tree. Under this sat a few mammies selling food. They sold fried fish and kassava bread, peppermint, rice pap, boiled or roasted ground-nuts (peanuts), corn (popcorn or Indian corn) and kayan (mendi or white), akara kuru and beans.

These local delicacies were displayed on wide calabashes with lids. The sellers were addressed by their wares. Thus there was 'mammy fried fish', as well as 'mammy akara', 'mammy kayan' and so on. All the foodstuffs were home-made and we had all sometime or another seen them prepared in our own homes. We each had our favourites, and invested our pennies accordingly.

Outside the gates was a larger assortment of sellers, but we were discouraged from going outside the gates during the mid-morning break. Not unnaturally, being placed out of bounds rendered their wares more attractive in our eyes, and their patience and strategy were well rewarded when the midday recess came, and we dashed out to spend our little pocket moneys on these suspended delights. In my case special arrangements for midday meals were made when Arthur joined me later.

The school compound was fenced in with iron railings. Against the south fence were the boys' toilets. These were called latrines, in the brutal terminology of colonial territories. There was a row of six cubicles, each with a bucket containing carbolic, which were emptied once or twice a day by prisoners, who drew them out through apertures at the back of the building. The building was of stone and corrugated roofing.

In front of it was the large pump of fresh drinking water, standing in a wide paved-in trough which drained away suitably. Needless to say this was where we spent most of our time. We had to be careful that the rush of water when we turned the handle did not wet our clothes, as this was certainly punished when we returned to our classes.

Most of the break was spent in eating, playing and wrestling. Kicking a tennis ball or a football was the most popular game, with cricket, racing, skipping, next in order.

Life was good, until we came to our first class examination at the end of that term. In arithmetic, reading, spelling, dictation, and moral instruction, I did not do so badly. But when the geography questions were put out on the board,

I could not make a thing of them. I forget now what they were but I do remember one question, and my answer. It was this:

Question: 'What is a basin?'

To this I put down the only answer I could think of:

'A basin is what you put soup in.'

I scored exactly zero in that geography paper. An all-time record in my life. It has always remained a mystery to me how one could score zero. It must be as difficult as it is to score 100, if not more difficult, just as metaphysically zero and infinity are impossible concepts (see Bertrand Russell's *Philosophical Theory of Mathematics*). It must be impossible, surely, completely to empty one's mind of all the facts one has learned? Be that as it may, I did manage it on this occasion!

Geography was always a peculiar subject with me. At the age of 11 we took the Preliminary Examination of the Cambridge University Local Examinations Syndicate. I passed in the examination, including Latin and algebra, but failed in geography. At the age of 13, when we took the Junior School Certificate Examination of the same university, I passed, including Greek, but again failed in geography. At the age of 15, when we sat for the Cambridge Senior, or the School Certificate, now the General Certificate of Education (ordinary level), geography was one of the four subjects I had Distinction in, plus two credits and two 'Very Goods'. The only difference I could see was that my father took me in hand for the last examination, and was in effect my geography teacher. But please do not think that I am absolving myself from blame, or imputing any to my teachers, whom I thank for all the hard work they put in on me.

★   ★   ★

However at school I improved with time, and soon I began to be included among the boys and girls who were called upon by the teacher to act as prefects.

I have said before that I was about the smallest boy in class, at least in size. I sat right in front and sometimes, I blush to say it, I shared the same desk with a girl. I mean I blush *now*; not *then*. The smallest boys and the girls sat in the front, and the desks and their occupants increased in size from in front backwards. Whenever the teacher left the class he would appoint one of the prefects to stand up and keep order.

The prefect stood in front of the class, by the side of the teacher's desk, and everybody did his own work. It was the prefect's job to see that there was no talking. Should there be, he called for silence and order. Usually it worked. But sometimes there was rebellion, and then he would issue a stern personal warning. Should this be ignored it became his 'painful duty' either to ask the culprit to stand up, and remain so standing until the teacher came, or more simply to write his name on the board. In either case the teacher punished the miscreant.

Thus there often developed what I can see now was a very funny situation. Picture a midget of a boy, that is myself, standing in front of a class, telling lads almost twice his size to shut up, stand up, or be whacked! I think I must have been very naïve.

It says much for the class that this situation hardly ever led to recrimination. I say hardly ever, because I can remember at least one occasion when nemesis reared its ugly head. We were often threatened, of course. And many a time when school closed, I would take great care to get ready quickly and make sure that I left school in the company of Mr Cole. But one day, somehow, I missed this convoy. And that was the day a certain big boy had lost his sense of humour. I knew his threat was not empty this time. Worried, I got ready, donned my hat, took my satchel, and was walking off briskly with some of the other boys who lived in my direction, when I heard someone shout my name. Did I say 'shout'? I meant 'roar'!

'Hi! Ageh! Wait for me!' the voice said.

What an optimist! To put it mildly, I did the opposite. And he started running after me.

Now this boy was not a boy, but a man. His name was Mohammed Bundukar, the largest boy in class. It was his job to mount us on his back when teacher chose that position for administering the cane. It was rumoured that he was married. And, looking back at the matter impartially, I am sure he was. He was a Mohammedan; and they married earlier than Christians. And he was in fact engaged in trade during vacations. He travelled extensively in pursuit of his business. He was a sign of that spirit of progressiveness which made a man in his position come to school to learn to read and write and generally better himself. Anyway he was furiously mad that day, and he chased me.

In such an emergency the Krios have a saying:

'*Fut we'tin a it a no gi yu?*'

Freely translated this means: 'Dear feet of mine, have I ever withheld anything from you? Have I not eaten so that you may be strong? Please do your job now and get me out of this mess!'

I live to record that my feet did not let me down. They responded with alacrity. Thus could be seen, on this tropical afternoon, a land race which, had it been on the high seas, might well have been between a galleon bearing down in full trim and a small pinnace trying to break away. The simile is rather apt as Mohammed B. was dressed in flowing Arabic garb, and I in shorts! In later years I could sympathise with the English over the business of the Spanish Armada.

I ran, I sped! He bore down on me, but just could not catch up. Soon I began to pull away from him, and then I noticed, with thankfulness, that people in the street, seeing what was happening, started to shout at him to leave the little boy alone!

'Bully!' they said.

'Yah, Bully!' I echoed in my heaving breast.

Then I had an idea. Instead of attempting to pull away from him for good, why not keep just out of reach in front, and let public opinion do the rest?

No sooner thought than done. It worked. More people took up the cry; and I noticed he began to slacken with irresolution. I could not resist the temptation to stop. This was the last straw. He leapt at me, missed, and in exasperation picked up some rubbish on the highway and shied it at me. Even that missed me. But it was his last gesture, and he left me alone as I turned the corner into Foulah Town.

The next morning when I arrived in class he shook his fist at me, but the matter was over.

13

Rehabilitation at Home

WE were a mixed class of pupils. We represented a very small percentage of the children who wanted to attend school. Education was neither compulsory, nor general, nor free,

СК

whether in Freetown or anywhere else in West Africa for that matter. Schools were very few. Children attended school at the age their parents could afford; and at school they were graded not by age but by educational development. Thus it was possible for a boy of seven to be in the same class as a boy of seventeen.

I have said that during the morning break, and luncheon recess, we spent the time eating, playing and fighting. We fought in different ways, on different occasions and from different causes. Ninety-nine per cent of it was good humoured, and consisted mainly of wrestling and shadow boxing. The difficulty about the latter was that in the heat of things you might forget to pull your punch, or your opponent might think you did so on purpose. A slight disagreement could easily lead to a fight.

Often even without any real difference of opinion, a fight would be started by the other boys. This was easily done. All you had to do was to get two boys together, and goad them to fight by gripping a tuft of the hair of one, and, transferring your hand to the head of the other boy, say:

'There! He has got your hair! Are you going to let him keep it?' I ask you: was the other going to let him get away with it?

The answer was 'no!' He jolly well was not going to let him get away with it. So he leapt at the other and grabbed at his hair, and hey, presto! the fight was on. The other boys immediately made a circle round the two, and they were off! Assembled opinion saw to it that there was fair play. And when it was over both shook hands, or did whatever was our equivalent of shaking hands. Then some other couple would probably take their place in the arena, eager to show their prowess.

4-2

Sometimes, however, these fights were genuine affairs, the result of real quarrels. When this happened they were usually settled after school hours. Real quarrelsome fighting during school hours in the playground was of course banned. Any boy who infringed this rule came inevitably before the principal, with equally inevitable results.

I seem to have had my share of such fighting, usually because of a facility in answering back, instead of giving the quiet answer that turneth away wrath. Usually these differences started as a cloud no bigger than a man's hand. Somehow they never seemed to occur in the morning on the way to school. We were in a hurry then. But on the way home after the day's work was over, with our minds relaxed and our hands itchy and empty, the devil had little difficulty in finding some mischief for our idle hands to do. I for one did not mind. At least it brought diversion to the journey home.

There were one or two empty pieces of land off Mountain Cut, where I have had my shirt soiled by the clinch of grimy hands or the dust of hard ground. At worst there might be a tell-tale tear. Even my mother got to know the signs. As we usually first removed our coats before the fight, the combination of clean intact coat and soiled or torn shirt and dusty pants usually gave the game away. As for father, I pass over his reactions tactfully.

Incidentally the fact that the fights occurred on the way home shows that they were disagreements between friends, namely boys who travelled together daily. They never reached the bloody stage, and were usually settled when one of us hit the ground sufficiently hard. I do not remember ever going home with a bruised eye or a cut face or my doing the same to another boy. Possibly we were not the tough guys we thought we were!

Even so, more often than not I received a thrashing for my pains when I got home. If from mother, this was because of the damage to my clothes. If from father, because of the damage to my morals. No one seemed to care about my morale.

One incident, however, nearly led to tragedy. There were in our class and the one next below, four West Indian children named Chandler, two boys and two girls. Their father was a sergeant in the West Indian regiment which in those days was stationed in Freetown. Three of them were what we Krios termed 'copper-coloured', that is the colour of a newly minted penny. The fourth and eldest, a girl, was extremely beautiful, and of that creamy complexion which, combined with blue eyes and brown or black hair, seems to be a speciality of the West Indian islands. I believe she was what they call 'octoroon'. We had no such colour scales in West Africa. There you were either black, white, or mulatto.

I have dwelt on this point simply because the Chandlers were the only West Indian children I knew in my childhood. Indeed, at the time they were the only 'foreigners' I had ever known, that is apart from Principal Holloway, Mrs Mavrogadarto, and the-one-that-got-away, my first white man of Malta Street. Bishop Walmsley, though English, was very much one of us, and did not seem a foreigner to us.

Horace Chandler and I were in the same class. We were of the same age and build. He wore boots; I went barefoot. We were quite pally. I forget what the dispute was about— probably something trivial—for I was surprised when we began to quarrel. Well, the usual dust-up would clear the air, no doubt. The usual crowd of boys accompanied us after school, anticipating fun.

Suddenly, as we came outside the gates, without warning,

Horace aimed a violent kick at me. What is more, he connected. It hurt. Immediately we grabbed hold of each
other and wrestled. Stung by the kick, I soon had him on the
ground, and according to the unwritten rules I had won, and
honour was served. I got up, and started to dust my clothes,
worried about what father would say, when a shout made me
turn round. Horace had an open clasp knife in his hand and
was coming after me!

I was frightened. I hated the sight of blood, or the thought
of being cut. Even today I cannot watch a boxing match in
the flesh, although I enjoy it on the television or cinema-
tograph screen.

That day, however, I was in a quandary. To run away was
unthinkable. In any case at the moment I was the adjudged
victor. But I was frightened; and dreadfully tensed. I dared
not take my eyes off the deadly weapon in his hand. My one
compulsive thought was to snatch it from him and use it on
him, before he could hurt me. The instinct of self-preserva-
tion turned me from a coward into a potential killer.

Fortunately the bigger boys promptly intervened and
kicked the thing from his hand, and threatened to thrash
him unless he behaved himself. We were never friends again.
He was altogether a different fellow from Mohammed
Bundukar....

But this was the only unsavoury incident of those halcyon
days. In general we continued to annoy, to tease, and to
quarrel, but always to retain our friendships and our loyalties.
For example I had not been at school very long when, at play
one day, a boy looked at me and said:

'*Folgot!*'

At first I did not get it. And then it dawned on me that he
was alluding to me, and making a rude comment about my

neck. The facetious fellow apparently saw in my neck a striking resemblance to that of a plucked hen! I hurled myself at him, and beat him hollow. He got up, and, walking away, held up his right forearm so that the wrist drooped gracefully like a hen's neck, the fingers spreading out like its comb. That did the trick! The other boys roared with laughter, and the name stuck.

I should of course have ignored the whole thing. But it hit me on the raw, with just that hint of truth which is the basis of all successful caricature. My neck *was* rather long. I was mad. I was fighting most of the time. This was easy. For, curiously enough, it was usually the smaller boys who took delight in baiting a bigger lad who had lost ground with the herd. I beat them. But they were hollow victories. Fortunately everything wears off in the end, and at last I was beginning to regain my old poise, as the nagging became more and more spasmodic. And then one day fickle Fortune made amends. It was the beginning of the next term when I saw in the playground a boy at the sight of whom I could only think of one thing!

'*Tolo-tolo!*' I yelled, raising my right hand in the now familiar gesture. Only this time it was a turkey I was indicating. *Tolo-tolo* is a word which reproduces the sound of the turkey's gobble admirably, and is the Krio name for the bird.

It was a masterly inspiration on my part. One look at the lad confirmed the aptness of the nickname even to the most obtuse boy in the quadrangle. At a stroke I had regained my position in the group. And my peace of mind.

★   ★   ★

At home, I was often in trouble for something not done, or more often for something done against repeated injunctions to the contrary. I was 'strongheaded', as father put it. The word means obstinate. There was a period when scarcely a week passed without some severe punishment from him. The punishment was applied in a wide variety of forms, from simply being sent to stand in a corner to the most severe caning.

If I had not known, even then, that my father loved me, I should have been horrified at the different forms of 'torture' he devised.

Take the simple case of standing in a corner. In the simplest form, reserved for the mildest offence, you simply stood, and let shame do its work. The psychological approach. For more serious offences you stood with both hands upraised and father sat behind you, usually reading. Should you let your tired arms droop, a sharp stroke of the cane on your bare calves brought you to your senses.

For somewhat more serious misdemeanours, in addition to standing with both arms raised you lifted one leg, and kept it lifted until the period of punishment was over.

If you were allowed to choose which leg to raise, naturally you stood on the right foot and raised the left. But you might just as easily be ordered to stand on your left leg and raise your right. Or to hold the lifted leg up in front, bent at the knee and hip, like the small letter 'h', or to lift it backwards like a man limping on crutches. The muscular fatigue varied with each position adopted.

Instead of keeping your eyes open, you might be ordered to shut them, this making the whole procedure considerably more difficult. In all these cases, father was trying to avoid caning us. He only did that as a last resort, when he was really vexed.

He was a man of inexhaustible patience. So far as the disciplinary side of the rehabilitation was concerned, he tried to vary the punishment for these monotonously recurring misdemeanours. By this means he probably saved his own sanity. So much so that sometimes we were actually grateful to receive a particular punishment, severe enough in itself, when we had expected something far worse.

It is only fair to say that I was far more often involved in these painful father-and-son sessions than my brothers and sisters. On the other hand it might well be that either father was getting tired of the unequal battle with each successive child, or else the stock of children was improving. If the latter was true, then the difference between me and my youngest brother and my sisters must have been extreme. I hardly ever remember them being chastised.

Another possible reason was that after the first three boys, one of whom died, the next two children were girls. This probably stumped father. For even an African father could hardly be as physically realistic with girls as with boys. Perhaps he was so relieved at the change of sex that he gratefully tendered his 'nunc dimittis'?

In addition to the forms of chastisement already described, he had other devices short of the rod. One of these was the punishment of 'picking pin', so called because you stooped as if to pick up a pin from the ground—except that you also had to lift up one leg behind you, the free arm draped along the body and the upraised leg. The arm which was picking the pin pointed straight to the floor, lightly touching it. You were not allowed to put any weight on it. The whole punishment consisted in maintaining this precarious balance, like a ballet dancer out of practice. If you failed you were beaten.

Whether these various 'advanced' methods of correction

were better or more understanding than the good old-fashioned cure of the rod is open to discussion. On balance, we the culprits often preferred the latter. It hurt more, but it was soon over. And in any case father was more helpful than he gave himself credit for, when he caned us.

A methodical man, he always had the self-discipline to restrict himself to a set number of strokes. The rub was that you never knew whether it was going to be three, six or twelve. If after three you received a fourth stroke, you waited patiently for the sixth. If another stroke followed this, then you knew it was a case for twelve strokes. We did have one practical guide. If, when the cane was produced, you were asked to stretch out your hand, you knew it was a minor affair of three or at most six strokes. If, on the other hand, one of the 'boys' (servants) was called in, this meant a mounting, usually a matter of a dozen of the best.

It may be thought by some readers that all this was harsh. Arthur and I certainly thought so at the time. So when the year after the war father proceeded to England, to study advances in water-works engineering and management, we welcomed his absence with joy. Mother, who had seldom figured hitherto in these barbarous rites, surely would be different!

Alas, we were doomed to disappointment. Mother took her duty so seriously that she mercilessly thrashed us on the slightest provocation. What was worse, unlike father, she went on and on, and only stopped when she was exhausted emotionally. Sometimes she would start again and beat us a second time for the same offence. There was no point in counting the strokes. While father selected the site of application and kept to it, mother just hit us on whatever part of our wriggling anatomy she could reach. Most of the strokes fell on our heads, elbows, neck and sides, and only a few on the really soft parts. We longed for the day of father's return.

I realise now that she must have been emotionally upset by father's absence abroad, the first major separation of their married life. At such a time the antics of two mettlesome sons must have been trying to her nerves.

The queer thing is we looked upon both our parents with the greatest pride in the world. I am only sorry we did not make their task easier.

★　　★　　★

In punishing us father managed to convey to us the disappointment and distress which he and mother felt at our behaviour. It was quite true that the caning 'hurt them more than it hurt us'. But he tried constantly to rehabilitate us.

Usually he would wait until the crying and emotional reaction was over, and then he would call me, or whichever child was the culprit, and talk quietly and earnestly to us. He always started with the same question:

'Now, Ageh.'

'Sir?'

'Do you know why you have been punished?'

The curious thing was we always knew why, and would answer:

'Yes, sir.'

'Why?' he would ask, and we would answer simply,

'Because I did so and so, sir.'

As far as I can recall, these severe chastisements were usually for some act of natural cussedness, which in any case would call for punishment, or else for some act of gross disobedience. If the former was the case he would ask:

'Why did you do it?' This question was not merely

rhetorical. And, surprisingly, such was the relationship between us that, if I felt I had reason for what I did, I would tell him frankly. And he would listen, and then explain why I was wrong. It was the same method he adopted in all things—getting us to think for ourselves, to know right from wrong, and to take the consequences of our action.

We were not frightened of our parents, and as a result we did not learn to lie. Father was very distressed if he thought we were lying, and this was to him an even more grievous sin.

On the other more frequent occasions when I had disobeyed him in doing something I should not have done he would ask:

'Did I not tell you not to do this thing?'

'Yes, sir.'

'Then why did you do it?'

'I don't know, sir.' This of course was inadmissible.

'You must know', he would insist. 'Why did you do it when I told you not to do it?' Sometimes I would be bold enough to suggest that I disagreed with him. Thus I might say:

'But, sir, he hit me first.'

He would not lose his temper. Patiently he would go over the ground again.

'What does it say in the Bible about turning the other cheek?' he would ask, and I would quote Our Lord's teaching on the subject. And he would ask about what He said about the meek, and I would reply: 'They that are meek shall inherit the earth, sir.' All these were quotations we children had been made to learn early in our lives.

Then he would bring more contemporary and pragmatic arguments to bear on the subject, asking me to repeat the verse:

'Let dogs delight...', and I would recite:

> Let dogs delight to bark and bite,
> For 'tis their nature to,
> But little children must not fight,
> God hath not made them so.

He would point out that it was wrong to hit anyone, however great the provocation. There was always constituted authority to see that justice was done. *On no account were we to take the law into our own hands.*

That was the crux of the whole matter. Our parents were doing their best to convert us from natural barbarians to orderly citizens of society. And that surely is the meaning of education. But it was patient uphill work.

These earnest discussions after a beating had their inevitable result. Like everything father did, they always happened; and soon we began to dread them even more than the beatings. For they were a direct appeal to the spark of decency which even the hardiest little boy had in him. And we were not so hardy either. Just naturally human.

We began to feel ashamed of ourselves for doing wrong; for hurting father and mother. Above all there began to grow in us a sneaking feeling that we were not growing up to be like our dad. This realisation had a jolting effect on me. For he was our hero. We began to try to please him. And he would have a word of encouragement for us when we had been specially good.

In time the physical beatings receded into the background, and we became more conscious of the love and understanding which surrounded us. Sometimes father would come into my room in the middle of the night.

'Ageh, are you awake?', he would ask softly.

'Yes, sir.'

'Now Ageh, your mother and I are very worried about you', he would say, and he would recall some particular thing I had done, or some trend he had noticed. And then he would talk it over with me, as man to man, or rather as father to son. Often, long before he was finished, I would be crying silently on my pillow.

These sessions continued long after I had left behind the age of chastisement. Instead, they became a bond between us, through which father communicated with me on many subjects, so that despite our relationship and ages we became friends. And as I grew older, if I had a problem, I would look out for a time when he seemed relaxed, and I would go and sit by him, and say,

'Papa, so and so...' and he would listen.

'Yes, Ageh...' and we would thrash out the point together. The friendship thus forged lasted throughout his life; and years after I had left home, when I was a student in Europe, I was able to turn to him for advice on the many problems which came up from time to time, with complete confidence.

He was not the only member of the family who favoured these midnight sessions. Granny Smart, our maternal grandmother, would often come and talk to me in the same way. This usually happened on the last night of my visit to her at her home in the village of Regent.

She would come and sit by my bed, and scratch my head affectionately and soothingly for quite a time, talking the while. Her words, however, were usually in the form of a benediction.

'God will bless you, *bobo* (little son)...the God of Abraham Isaac and Jacob will bless you...be good; hate no one, harm no one...God go with you....'

My mother did not engage in these nocturnal talks. For

one thing she was rather shy and reserved with us. For another she was somewhat deaf, a family ailment from which she and her brothers and sister all suffered. Another family feature was a tendency to premature greyness. One of my maternal cousins turned grey while we were both at school.

Mother would get us to pull out the few strands of grey hair as they appeared among her locks, which were silken in texture. She was a beautiful woman, despite the fact that she had a bad attack of smallpox in childhood. While I sought out the errant grey strands she would talk to me. Not long discourses, but penetrating observations on life, people, our family, and us children.

Father, mother, grandmothers, are all gone. But they worked hard on us in this life, and their prayers still enfold us.

❧ 14. ❧

The War Years

I HAVE been asked how it is I can remember what happened so long ago in my life. The answer is simple really. There are certain landmarks which are easy to fix, and by which personal events could be related in time and place.

Take the case of Mohammed Bundukar, who, I have shown, chased me with murder in his heart because I had taken disciplinary action against him in class. I can picture the scene as he rose up in the full majesty of his years and robes to threaten me. This was in the classroom which adjoined and faced the Infants Department. The only period I was in that classroom was during my first year at school. Therefore I must have been seven years old at the time.

Another landmark which helped to fix events was the First World War. It is a coincidence that my elementary or primary school career began and ended in the same years as the war. I started primary school in January 1914 and finished in December 1918.

Curiously enough, I remember very little about that war. One incident I do distinctly remember, however, was going on an errand, as I often did in those days, to buy local food-stuffs for mother, and finding that the price had been more than trebled overnight. These foods included fish, rice, and *foo-foo*, a starchy foodstuff made from the flour of the kassava tuber. All were produced locally.

The previous day father had come home and announced, 'War has broken out in Europe'. To our queries he had explained to us children that the war was in a very, very far country. I was surprised therefore when the balls of *foo-foo* which used to be sold at two a halfpenny or thereabouts only the day before had gone up four-fold in price the very next day.

So I challenged the new price; but the food sellers replied:

'Have you not heard there is war?'

'Yes, but it is in Europe', I replied.

'Europe, Europe!' they mocked. 'Have you been to Europe?'

'I have not, but it is very far away, while this *foo-foo* and rice come from "up the line"' (that is, were produced in the interior of our own country). When father returned home I asked him why the prices of locally grown foodstuffs should suddenly rise just because war had broken out so far away; but all he could say was: 'That is what happens when war breaks out. All prices go up, and people try to make a lot of money.'

Apart from this, however, the war made no impact on my childhood. Certainly the armistice caused no ripple in our life. I cannot recall celebrations or jubilation or other outward signs.

I remember, however, father coming home one day and saying with quiet pride to mother: 'The Navy have written to the city council thanking them for the high quality of the water supplied to their ships and for the service given them.'

Throughout the war, Freetown was an important watering station for the British Royal Navy, and an important station of the Atlantic Fleet. The water was obtained from the Freetown Water Works, for which my father was responsible. It was about this time that I first heard the statement made that this our Bay on which Freetown stands was large enough to contain the entire British Fleet at anchor.

As to the British Army, we saw plenty of them at school. Exactly outside the main gates of the Model School, where Circular Road crossed the road connecting the main barracks of Tower Hill on the one hand and those of Mount Aureol and Kortright on the other, British soldiers could be seen daily, marching to and from either barracks, sometimes in formation, sometimes in small groups, often in pairs or singly.

They were called '*Jek-Jeks*'. They wore khaki shorts and shirts, heavy black hob-nailed boots and puttees.

Exactly opposite our main gate was a house where some young women lived, a typical wooden Freetown house. This house was a regular port of call for the *Jek-Jeks*. Often they would be buttoning or hitching their trousers or fastening their belts, or combing their hair, as they came out.

In our innocence we took all this with wide-open eyes, listening half comprehendingly to the comments of the bigger boys.

As I look back now on this scene it is curious that the authorities saw no incongruity in a house of prostitutes at the main gates of the Government's show school, where the new generation of African youth was being given their first steps in a liberal education.

*   *   *

In all my five years at this primary school my third year stands out as a watershed, by which in retrospect I can fix such events as I can remember. This was the year in which I was nine.

The year itself started with a Narrow Escape. An event which in the light of modern views on child psychology, undoubtedly had a bearing on my development at this tender age.

My first two years at school had been fairly normal. My teacher in Class I, Mr Erasmus Cole, although a strict disciplinarian, was to us children an ordinary human being of a master. He was a good teacher and a strict but fair disciplinarian. We knew when to expect punishment and what to expect.

Our teacher in Class II, the next year, was Mr W. E. D. Campbell. Although he could use the cane when necessity arose, his general effect on us was that of a kindly and lovable teacher, the sort that boys have no hesitation in going up to speak to. In the corridor or in the street he would stop and talk to us, calling us by our Christian names.

He took us in groups to the Post Office Savings Bank and taught us to open savings accounts. He took us on rambles, and made nature study real for us. He planted beans in a bottle and showed us the wonder of roots, stalk, and leaves breaking forth into life as the bean became a plant. He even had a small museum in the form of a glass locker in which were specimen jars. One of these contained a small adder, coiled S-shaped so that we could see and identify the various parts of the body. He took us in singing in class, and in the whole school. In short he did all those things which transformed school from routine work to an exciting pastime.

We soon discovered which was his deaf ear, and in times of crisis we took care not to be caught on the 'wrong' side, that is, the side of the good ear!

It will be evident that we were happy under him. So when our year with him drew to its close, we were without exception heartbroken. There was a double reason for this. We were not only sorry to be leaving him, but we were appalled at the thought of what lay in front of us in the next class. The teacher in the next class was a gentleman who, in a school where all the masters were strict disciplinarians, and members of the Honourable Society of Kid Wallopers, had a special reputation for being positively awesome. He was a tall, slightly stooping, morose man, who never smiled. He had a habit of looking at us sideways, appraisingly, in a way which made us tremble.

Sounds of violent floggings from his class were frequent, and the stories circulated by boys in his class were hair-raising. He was said never to cane his pupils on the hand, but always mounted on someone's back. He was reputed even to do the same with girls. In later years he took Holy Orders.

I quailed at the thought of moving up to his form. And that is saying a lot, because my own father was one of the greatest caners of boys that I knew; but I was not frightened of *him*. On the contrary I was definitely frightened of the master in charge of Class III. When school resumed that January I felt ill. Unfortunately my temperature remained normal, and there were no confirmatory signs of my illness. So the following day, father accompanied me personally to school.

I need not have worried. The guardian angel of frightened little boys must have been active on our behalf. I found that there had been a switch among the masters and Mr Campbell had moved up with us to Class III, while the Terrible Master was now in charge of Class II. I never sat under him all those five years.

This year, which started so propitiously for me, was marked out in a number of other ways. This was the year for example when all the pupils of the school were measured for height. This was done with a six-foot rule fixed to the wall in the corridor. When the results were announced I was four feet four inches high. What was more, when the whole school had been measured Mr Campbell said to me:

'Ageh, do you know that you are the second smallest boy in the whole school?'

The shortest boy, also in our class, was called Mohammed Bokari. His height was four feet three inches.

This too was the year in which I started manual training. This addition to the curriculum was introduced when boys reached Class III. This was the only colony primary school in which it was provided. None of the mission schools did. On a certain morning each week the boys of our class spent the whole morning in the manual training department, while the girls went to have lessons in sewing and domestic science.

The manual training was in a separate building on the

south side of the compound, and we looked forward immensely to this new variation from book work. We soon mastered the use of hammer, plane, saw, chisel, T-square, and we were set on to make simple things. Our instructors were Mr Sylvanus Bull and Mr Reginald King. It is to this period of training that I owe what little I am able to understand about simple jobs in the home. Without it I should have remained a bookworm, clumsy in anything which called for the use of the hands.

We were a mettlesome group when we went to manual training, and Mr Bull and his staff often had difficulty in keeping the class in order. Now and again, when things got really out of hand, he would collect a group of boys and march them off to the European principal. The latter's treatment was simple. Cane the whole lot! Then back they returned to manual training, little heroes in the eyes of the other boys!

One day, in the commotion, an instrument was broken. I forget now which it was exactly. But it was really serious this time. In the inevitable round-up I was included, although I was not implicated. I merely happened to be standing by the nearest bench when Mr Bull made his selection: 'You, you, you, you, you, and you.' The last 'you' was I.

Never backward when I was innocent, I protested my innocence all the way to the principal. I continued to protest to that gentleman, who, I was sure, would be fair and at least would listen to my plea of innocence. The other lads remained silent.

The principal listened to me. Or rather he listened at me. Then he dismissed all the others, without even questioning them, and without a word he thrashed me mercilessly! Not a word was said through all this by him. Then he dismissed me.

As for me, I refused to cry, although it hurt terribly.

It was in this year too that I joined the choir at our church. Ever since I started attending church I had wanted to become a chorister. In fact in after years, father used my love for the choir as an aid to punishment. In cases of grave misdemeanour he would stop me from going to robe for one or two Sundays according to the severity of the penance.

Now at last in my ninth year I was allowed to go up and see our organist Mr Joe Metzger. He gave me a trial, and I was allowed to robe only two Sundays afterwards, a record period of probation at the time. But then I had had long practice, singing at family prayers at home. After this, for the next nineteen years of my life I remained a chorister, twelve years in my native Freetown, and seven years while a student in England, including an oratorio at the Crystal Palace, London.

Later, also during this my ninth year, I started music lessons with Mr Metzger; and this together with my choir activities opened up a new and pleasant side of my life.

Two further incidents occurred in that year, which have had a bearing on my future career. The first was a particularly severe caning from my father. It came about in this way. In my first year at school in Class I, in a class of about thirty, at the terminal examinations I was twelfth at the end of the first term, tenth the second term, and first the third term. And in the total for the whole year I was first. These marks were always read out to the assembled class. In the next class I was first in the first term, second the second term, and third the third term, and first for the whole year.

Then we moved to Class III, and I was fifth at the end of the first term. Father was extremely annoyed and gave me the thrashing of my life, saying that God had given me brains and he was not going to stand by and see me waste them; in future whenever I came lower than second in an examination I would be thrashed. I am thankful to say that there was never afterwards any need to put this threat into effect, whether at primary or grammar school, college, or later at university in England.

I was not surprised that he had reacted so violently on this occasion. I was well aware that I had been playful all that term. And at the terminal examination I had prayed extra fervently that God would help me. But in my heart I knew I did not deserve His help, and was not surprised at the result.

My father was strict in many ways. He had very high moral standards and two things he was always waging war against were what he called my tendency to be a bully, and my pride. He was always preaching humility to us. And he was himself essentially a humble man. His motto for us children was always to make the most of our abilities, and at the same time to be humble. He often referred us to the parable of the Talents in the New Testament, and also to the story of the Tortoise and the Hare.

The second incident concerned the question of my career in life. At our terminal examinations that year we were asked to write an essay on what we would like to do when we grew up. This was a time when I was very serious-minded, and wanted to be a minister like our pastor. I saw in our church, our choir, our pastor, my highest and noblest ambition.

With all the urgency and excitement of a nine-year-old boy I poured out this my ambition in my essay to such good purpose that Mr Campbell was deeply impressed. He gave me ninety-four for my essay and I was first in the whole examination that term. Ever since then, for years afterwards, whenever he met me in the street, he would stop me and ask how I was getting on with my ambition to become a minister.

I have not become a minister of religion; and the reason is a familiar one. As I grew up, even while I was still at school, I became more and more aware of my shortcomings. The very efforts of my father to make me a good boy and a decent citizen, and the earnestness with which he pointed out my failings, in time began to convince me that I was far from perfect. And, in the honest eyes of a child, a minister was the very essence of goodness and perfection.

In my childhood's eyes the surplice was the physical embodiment of the phrase: 'White as snow.' I knew in my own red-blooded heart that my soul was not white as snow.

Gradually I began to lose confidence in *myself* as a candidate for this highest calling. By the time I had gone to secondary school and college, the disillusionment was complete, and I had begun to look around for another calling, which I could pursue as an ordinary human being, without the high ethical standards which the ministry called for.

Something else happened in this fateful ninth year of my life, when I was in Class III, which helps me to assess my progress at this period, and throws some light on my life then. It was this year that father asked mother to cut out all extra domestic work from my programme. From then on, I was to concentrate on serious study. Mother readily acquiesced. She had done her job training me to be useful in the house, and in any case there were many servants and children available.

I was then only nine years of age. And yet by that time I could cook many of our African dishes. I had been trained to go shopping for the needs of the house, especially foodstuffs. I was already doing this when war broke out and I was seven. Mother preferred to send her own children rather than house boys. She was deliberately training us. At first we would accompany one of the older foster children, then later she would send a house boy with us for protection.

I loved to go on these shopping errands, and it was not long before I found out that if (say) I was told the price of a piece of fish was two shillings, it was policy to offer (say) six-pence. In the arguing which ensued a fair enough price was usually agreed on in the end. Most of these food sellers knew who my parents were, and put up the price accordingly. At times, however, when I so blatantly undercut the price, a fisher-woman would get really angry and brandish the fish at me, shouting: 'Get away you *Bobo* Cole, or I will slap you with this fish.' But when I made to move off she would call me back.

At weekends I would go out selling for my paternal grandmother, or my mother.

Granny Cole sold freshly ground coffee and chocolate. She got the beans from the village of Regent, either from or through granny Smart my maternal grandmother. These we would husk, dry in the sun, and parch in a red-hot iron pot. On Friday evenings she would roast and grind the coffee. Actually we pounded it with a wooden pestle and mortar, sifting it through a fine-mesh wire sieve, pounding the coarse grains and repeating the process until the whole had been reduced into a fragrant fine black powder, as fine as flour.

This was made up into little bundles and stacked in a flat calabash which I would carry on my head and set forth, down Pownall Street, along Fourah Bay Road, towards the centre of town, selling with great aplomb. I did not have far to go, because along that route my grandmother's coffee was popular among most Krio families. I never liked to return home with-out having sold every single packet. Should I have a few left, I would go further afield hawking around until I had sold the whole consignment.

Sometimes I went out selling home-made drinking chocolate in slabs, sometimes lettuce. This last we used to get from Regent and the other mountain villages. It was fresh and crisp, and was very soon sold out.

These were the activities father stopped when I reached the age of nine.

In the morning we had to take turn in tidying up the house; and it was my job to tidy up the parlour and mother's room, dust her dressing-table and sweep up. We had to clean father's boots, and fold and put away his suits.

It was also my job to 'wash plate', that is, to do the dishes before going to school. They were stacked in a large metal container and taken to the pump in the backyard. There, with soap and water I would wash and rinse them, and then take them to the kitchen.

Obviously these various activities did not all take place at the same time, but at one time or another they were an integral part of my daily life. Washing the dishes was often the cause of my being late for school. I had ample time to do them, but I was a dreamer by nature, and, instead of getting on with the job, I would stand or sit with one arm tucked behind me and with the other I would dangle the dishcloth in the water, and run my hand round and round, idly watching the ripples. Thus I would stay lost to the world and my task, when father's shout would wake me to reality. Sometimes, at the very end of his tether, he would beat me to hurry me on with the work. But it did not change my nature, which has remained essentially reflective.

That something has been made of my scholastic life is due very largely to the close and careful supervision which my father paid to everything connected with my school work. There was never a day when, however tired, on returning home he would not ask me what I had done at school. He would listen carefully to my report and sometimes would question me on it. He would look at our books and sometimes set us simple tests.

We had the feeling that not only was he keeping a check on our work, but he was genuinely interested in us and what we did. Not unnaturally we did not always relish this close interest in our work, and often my brother and I would hope that he might miss a day. But this rarely happened. On an occasion when we had gone to bed, without going through this catechism, I have known him come to our room, and if we had not yet fallen asleep, ask us the inevitable question. Sometimes it was on the following morning, or even the following evening.

Years afterwards, when I was at Fourah Bay College, I was complaining to the other students about this scrutiny from which escape was impossible, when another student said wistfully:

'I wish my dad took half the same interest in me.'

I was very much ashamed after this, and it has helped me to appreciate what my father, and my mother, did for me and my brothers and sisters.

One further important event occurred in this fateful ninth year. My second sister Irene was born. She was the last of us to be born in Pownall Street, Kossoh Town.

Fleshpots

BY this time I was no longer a child but was now a boy, with all the problems this stage of development involved. My faculties were sharpening and I was reacting fully to my surroundings. As a result I was beginning to be a problem to my parents. Getting wild, in fact.

In Kossoh Town we were surrounded by the fleshpots of Native Africa.

Immediately in front of us in the house of Mammy Sama there lived some dozen Limba people, mostly adults. They were *woroko-worokos* (illiterate pagan labourers). As soon as evening came, the men pulled out their low stools or logs and sat in front of the house playing *warri*. This was a game played with cowrie shells and an oblong wooden log which had a number of cups scooped out in double rows. It was a popular and absorbing game, and friends and passers-by would stand around watching.

Then later, just when it was about time for us children to go to bed, the real fun would begin. A musician or two would arrive with *balanjis* round their necks. These *balanjis* are the African prototypes of the xylophone. They consisted of a keyboard of special hard wood of different lengths, like the keys of the xylophone, with gourds tied underneath to provide resonance. The keys were beaten with solid club-headed sticks. The musicians had little bells attached to their wrists to impart a tinkling overtone to the music.

Other musicians would join in, some with zither-like instruments, others with stringed instruments which they bowed or else plucked. As they played, the dancing would commence, until at times a full dance would go on all evening, the men by themselves in closed crocodile formation, the women by themselves, rarely the two together. This dance is very much like the conga.

A little lower down the street from us was the compound of Daddy Ali, the Moslem Timne chief, in whose compound big gatherings and festivities were frequently held. These reached a climax during the Ramadan season.

I learnt about the Ramadan from a number of my class-mates at school who were Moslems. During its forty days of fasting they used to carry empty cigarette tins in which batches of fifty cigarettes were imported. These they placed under their desks; and from time to time they would draw them out and spit into them with gusto. They were not allowed to swallow even their spittle during the fast.

One day the boy with whom I shared a desk, a Moslem, was looking in his pocket for certain things when out came a metal spoon.

'What is that for?' I asked. He was my pal, so he answered me frankly:

'This is for the feasting', he said.

'Feasting?' I asked. 'I thought you were fasting?'

'Oh yes, we fast all day, but we eat before sunrise and after sunset. We eat as much as we can then.'

'You forgot to leave your spoon at home then?' I asked, puzzled.

He smiled. 'You mean I remembered to take it with me wherever I go so as to have it ready when fast is over. I may be away from home then. Any faithful *Marabu* (Moslem) can eat anywhere there are *Marabus* feasting', he explained, 'so long as he first prays.'

At Chief Ali's compound there was always a huge throng of people tucking in at the feasts held every evening for the faithful.

The end of the Ramadan season was celebrated with mammoth feasting and a carnival procession of magic lanterns and paper objects in the form of ships, railways, giant dolls, and charades.

At Easton Street, the next street behind us, was the compound of a Santigi or minor headman of the Timnes. Unlike Chief Ali he was not a Moslem, but a pagan. As such his was the centre of many native Secret Societies and their attendant festivities. *Wende*, *Poro*, *Bundu*, *Alikali*, and other 'devils' paraded there, along Malta Street, and down Pownall Street.

We were strictly forbidden to join the throng on the street, still less to watch these festivities at the Santigi's compound. But that made little difference, for all we had to do was to remain obediently upstairs in our own home, and look through the open window as they passed along the street below us. A procession of *Bundu* dancing girls, their bodies shining with ceremonial anointing, would 'come out' from the bundu bush where they had undergone their initiation rites, and were now ready to marry and take their place in tribal society. They were well developed and attractive girls with all the outward signs of budding womanhood. At the head of the dancing procession was the *Bundu* 'devil', clothed in a grass skirt, and carved black wooden mask.

One day, in one of these many processions, we had the thrill of our young lives when we saw two *mamampara* men. We were leaning out of our upstairs window, taking in the spectacle, when the men appeared, and were so tall that they looked straight at us and past us into the parlour behind us. They were clowns on stilts, and we were so thrilled that for

weeks we too tried to see if we could walk on stilts, using fencing poles for the purpose.

Then there were the *Egugu* devils; the *Poro*, *Alikali*, and other gods and minor deities with their devotees, all of whom provided a spectacle greatly attractive to children. Unfortunately they were all heathen or anti-Christian.

Kossoh Town was, and still is, an enclave, surrounded by Patton Street, Kissy Road, Savage Square, and Fourah Bay Road. Along these thoroughfares the main traffic passed, leaving Kossoh Town and its people to their practices and customs in isolated independence.

Immediately to the east of us, in fact starting from Easton Street, the very next street to ours, and stretching eastwards to Savage Square, was the small district of Krojimi, which might be termed the 'Deep South' of Freetown. Here people of old Yoruba stock lived, untouched equally by Christianity or Mohammedanism. Paths led in to it from three sides but faded into compounds and areas which led one to another. There were no streets and we were never allowed to go there.

From west to east, Kossoh Town, Krojimi and Fourah Bay were, broadly speaking, inhabited by people of Christian, pagan and Mohammedan Yoruba stock respectively. In addition Kossoh Town was heavily impregnated by pagan natives from the interior of Sierra Leone. These were mainly Timnes but included a few Limbas.

I can see now how urgent must have been the dilemma in which father found himself, in his effort to bring up his young family in the best conditions. A strict orthodox Christian himself, he was surrounded on all sides by pagans and Moslems, of both Sierra Leonean and Nigerian stock. His children, decently clothed, were the boon companions of children who went naked even in puberty. Apart from this, these conditions interfered with that routine which is so essential in growing schoolchildren. That is, regular time spent on homework, early to bed, and restful sleep.

We were always suffering from fever. These bouts of malaria lasted two or three days and left us weak and anaemic. Keenly aware himself of the measures necessary to combat the breeding of mosquitoes, father was surrounded by people whose public health sense was nil.

These must have been some of the problems which confronted him. But with that foresight which he displayed in all his dealings with his family, he had long ago set in motion certain plans. These were to prove decisive at this psychological period.

❧ 16 ❧

Eden Discovered

ONE day, some time previously, father had asked me if I would like a ride in the hammock. This was a treat reserved for special occasions. Usually it meant a visit to granny Smart and Regent, so I perked up with expectation.

Regent, and the lovely sugar canes! And uncle Richard with the bushiest and shaggiest of whiskers, and such a huge beard, with a little space inside for a face! And yet his voice was so gentle, and his eyes were always twinkling in a crow's nest of a smile.

The sugar canes came right up to his very doorstep. And, final touch of fairyland, the stream along which they grew actually flowed under the house, the pillars of the latter rising astride it. To complete the picture there were orange groves, guava trees, paw paws, pineapples and plum, and a small corner where he grew lettuce and shallots for the Freetown market. There was always a nice fragrance about the house. And we never called on him without his going to the garden, with a shining cutlass, and cutting us a long juicy cane twice my height.

So when this day father asked if I would like a ride on the hammock, it was like saying: 'Would you like to go on a picnic, or a trip to the seaside?' Would I? I certainly would!

I hurried off to dress. I did not have to be told to wash my hands and my feet. But I was told to put on my boots. This must be a special occasion! And I wondered why my parents kept looking at me. They seemed somehow younger, happier, almost like grown-up brother and sister to me.

I had been ill for some time and I had been kept indoors for what seemed a long time. The hammock was brought out. The four hammock boys took their places at the corners of the canopy, their little round cushioning *katas* on their heads. Mother took her seat, and placed me on her lap. The men moved off in perfect unison, face forwards, father leading the way on foot.

Right, left; right, left; I nestled in mother's comforting bosom, and made believe I was the principal of the school. I had then not long started school, and the novelty of seeing him astride his hammock was still strong. He was so different from father. I swelled up, puffed out my chest, patted my hat, pretending it was a pith helmet, thrust my head and neck forward, scowled slightly, blew out my cheeks and cocked my arms akimbo, with shoulders hunched high. Thus I rode that day. I was happy.

Every now and again the men would lift their burden—in unison—shake their heads and stretch their compressed necks, and then replace their load, continuing with the jerky-sidey-hippy movement of persons carrying a heavy load. We must have gone a long way, for the next thing I remember we seemed to be in another part of the world altogether.

We were in a forest, dark, lofty, and cool, with gently rustling leaves. Large tree trunks, broader than father, shot up skywards. But it was the carpet of leaves I remember. Thick brown, soft, crumbly, downy leaves; and trees which I had not seen before. From their branches hung large bell-shaped fruit, jutting from the broad bottom of which were grey, smooth stones that looked like tiny men's heads.

'Cashew!' father said, and he picked a few red and deep yellow ones. He gently pressed his finger into them and the juice burst out. He gave mother one and me one. He took a bite, kept his teeth in and sucked. I took a bite, kept my teeth in and sucked! Delicious heavenly fruit.

'Do you like it?' my father asked.

'Oh yes, sir!' I replied.

We continued to walk on, father and mother on either side of me, each holding my hand. Presently father took out his surveyor's measuring tape and with the help of the boys spent quite a time measuring.

Once he shouted sharply to the boy in front, 'Look out.' The other turned round calmly, as if he had looked out long before father shouted. And I saw a long flat round thing moving away in the grass, S-shaped. It had no legs; it moved silently. I clutched my parents' hands.

We continued to climb and soon the most wonderful sight met my eyes. Below me, like giant green umbrellas, as far as the eyes could see, were the leafy tops of trees—mostly green and yellow ones. Some of the trees were short, some tall, and here and there a giant cotton tree topped all the others disdainfully.

The view stretched as far as my childhood's eye could reach, past more trees, rather bluer than green, dotted increasingly with what I realised were the tops of houses.

The land stopped suddenly and raggedly. Beyond it stretched a vast greyish blue expanse scattered all over with little white crests which seemed to be trying to reach the land. There were also little objects carrying what looked like yellow tablecloths.

'Do you see the sea?' my father asked.

'Yes, sir', I answered. I had never seen anything like this before. I had been to the seashore at Christmas time, when we went to collect sand for the church. But never did I realise it was so big.

But there was a further surprise for me. 'Let's go this way', father said, and he continued to lead us further into the wood. I noticed that the grass was greener. The air smelt fresher, damper. And, suddenly, we came upon some glistening rocks and a ten-foot waterfall tumbling into a pool!

'How do you like it?' father asked. For once words failed me.

'How would you like to live here?' he continued. It was too wonderful to be true. It was a veritable garden of Eden. The original Garden could not have been more idyllic. Nothing could be more pleasant than to live in this fairyland.

Papa had just bought the land, and as a special treat he had taken mother and me for a picnic there.

We were coming back, the 'head boy' leading, when again father shouted to him to be careful. But this time he was too late. For as the tell-tale slimy S-shaped thing some two feet long slid away into the undergrowth the man lifted up his foot and grabbed it hard. He stood stock still thus, like a heron, for several minutes, perfectly still, while my father chased the snake, throwing stones at it, and my mother kept telling him to be careful.

Then my father came back to hustle the man away to hospital. But the latter just continued to stand there, as if lost to the world....

And then, calmly, as if nothing had happened, he continued to lead the way back to where the other three men were waiting with the hammock. My father would not let him carry the load but ordered him to run to the hospital with one of the other men. But he refused. All he said was:

'Master, the snake, not me, will die!'

And that was that. He was not even ill. But the incident had a profound effect on the other three 'boys'. From that day father never again had to shout at them in desperation. Once his orders were passed on to them by the senior boy, they were obeyed promptly.

★   ★   ★

There then followed one of the happiest periods of my childhood that I can remember. There was quite an uplift of spirit at our home, with constant references to 'the farm'. Every Saturday father would come home early, and after a quick lunch would go out to the new property with the servants. Often he would take one or both of us children with him. Sometimes when he was specially stirred he would go there direct from work, and send the hammock for us.

I soon got to know the route. It was uphill all the way. Past Bethel, right up the whole length of Easton Street, across the main east-west Kissy Road, and along Upper

Easton Street in a graceful curve for almost a mile, passing two streams on the way.

It was in the first of these, called *Bushwata*, a spring gushing out of the solid rock into a sylvan bowl, that I first saw a woman bathing. She was a girl, really, in the full flush of womanhood. An African nymph, her twin breasts shapely beyond compare, firm, dancing up and down with the motions of her arms, as she scrubbed and poured the crystal water over her shining brown body. Over her hips she wore a tiara, six inches high, of large black beads, embracing the form in caressing intimacy, setting it forth to advantage.

Later I knew that this tiara of beads, worn by girls of the indigenous tribes, was called *jigida*. Our own Krio girls did not wear them, but they too were just as shapely and firm. I got to know all this with the passage of time. For that spring remained, throughout, the pool of naiads, though desecrated often enough by men bathing there, equally openly, equally naked.

The other stream was more prosaic. It came down from Mount Aureol somewhere; and all the road did was to cross it; without a glance. I believe it disappeared to the left through the foot of Merriman's farm. A public water standpipe stood near here, the highest in Freetown. It was exactly on a level with the main reservoir at Tower Hill some four miles away, and in times of drought was usually the first to dry up. Father never passed it without giving it a special glance. Probably if he were alone he would stop and pat it. He seemed to think that it was proof of his prowess as a water engineer.

He was also a brilliant mathematician, and knew when he had reached the limit of possibility. His farm stood much higher than this tap. His first step, therefore, was to build a dam at the home stream, erect a large galvanised iron cistern six feet across and high, and feed a pipe from the dam to the latter. We then had all the water we needed.

He went further. Also direct from the stream he ran feeder pipes at parallel distances all over the four-acre land. These pipes were mounted some three feet from the ground, on roller supports, with a lever at the end of each row, by which the whole length of pipe could be swivelled over a complete arc.

He kept this a special treat for us. One Saturday the whole family, including granny Cole, went to the farm. After a good look round, father assembled us near these long metal lines, and turned a stop-cock. Immediately fairyland broke loose. Along each row of pipes at every foot along its whole length, a graceful stream of water shot forth, at first thin, then bursting out into a little tumbling spray of a waterfall. Nipples had been fitted into the pipes.

Father took hold of a lever and swivelled one pipe. The direction of the sprays followed the movement of his arm from well over on the right, along a graceful arc, over to well down on the left. Immediately I ran to the next line and moved the lever. My cascade described a similar arc to father's, although more slowly. It was stiff. Arthur, not to be outdone, ran to the next. The same miracle happened. Slower, stiffer.

Soon everyone was at it. Mother, granny, there was a row for each. We started playing tunes, moving the cascades here and there in different rhythms, the resultant pattern a symphony in liquid crystal, like a pleasant display of fireworks drenched in snow.

The wind took a hand in the gambol, swishing suddenly and mischievously into the flying waters, and drenching us. We laughed. It laughed. The fountains chuckled.

'Well, what do you think of it?' father asked and mother kissed him.

'Why, it's wonderful!' she cried. She was like a little girl. We were all so excited.

'It is to water the plants', father explained. 'We shall be able to have rows and rows of fresh vegetables. With God's blessing we shall be happy here.'

He spoke prophetically. Years after, a young man came to buy fresh vegetables for his mother. He was a university graduate, an assistant master at the Grammar School. He was quiet, very well mannered. My parents liked him on sight. He liked my eldest sister on sight. And that was the first romance of the new generation at the farm.

Here father built our new home, with the aid of our maternal uncle Francis. A three-storied building of granite and ferro-concrete, the first privately owned house of that construction in the country. It was built on a broad artificially constructed terrace, which provided spacious enough ground for our boys' games, including home cricket and football.

The house was an advance on 13 Pownall Street in another direction. The kitchen was inside the house, on the ground floor. The fireplace was not stones and tripods, but a brand new Aga-type cooker, imported from America, with four apertures for cooking, a hot water reservoir and large oven. A special metal chimney was connected to the back of the flue and passed out up the side of the house to project above the roof. This must have been the first domestic chimney in a Sierra Leone house.

The day came when the house was to be christened. Pastor and several members from the church, and many grannies and friends from Kossoh Town, came up for the day. We had a regular *awujo* (ceremonial feasting). Everybody blessed the house. God had inspired father to build the house. God would surely bless it. The members of the family who had gone before to the other side of the grave would bless it.

A curious thing happened later that day. After all the feasting was over, and the neighbours and guests had gone, I was given a pile of crockery to wash. I had done this, and was taking them into the house when I fell down heavily backwards on my bottom, bumping the pile of dishes on the concrete floor in front of me. They split from top to bottom. But mother was not annoyed. She took it as a gesture from our ancestors welcoming us, and accepting it as our homage on coming to the new house.

❧ 17 ❧

Sunrise on Mount Aureol

SO we moved from Kossoh Town and went to live on the slopes of Mount Aureol, overlooking the east end of Freetown, and the Bay of Sierra Leone. I would be about ten then. My second sister Irene had been born, but not my third brother Eric, nor the last of the line, my sister Taiwo.

Nothing could be more dissimilar than the old life in Kossoh Town and the one which opened up in our new home. Down in town I never saw the sun rise nor really set, thanks to the fact that we were on level ground, and were surrounded by the trees and houses of the city. High up here there was nothing to obstruct the view across the eastern half of the city and the Bay, to the far-off shores of Bullom and Tassoh

Islands. High above everything there was nothing but blue skies down to the far horizon. At night I often lay on a table on the verandah and gazed at the stars, which were marvellously bright. And when the moon was out, its brilliance on the land and sea was almost overpowering.

We saw the sun rise every morning; first sending out purple feelers, then breaking forth in all the majesty of cataclysmic reds, crimsons, and gold, so penetrating and blinding, that no wonder man through the ages instinctively has bowed to it in worship.

The air was free from the dust and dirt stirred up in town by innumerable feet padding constantly throughout the day. We were seldom sick. The land sloped away and drained off all stagnant water in which mosquitoes could breed; and father could keep a wary eye on our own public health problems, such as refuse, empty tins, and the like.

We saw the ships riding at anchor in the bay, and turning round with the tide. We saw the native Bullom boats sailing down the river in the morning, bringing wood from the interior, together with such foodstuffs as dried fish, *foo-foo*, rice, kassava, yam. In the late afternoon they returned home, their graceful yellow inverted sails looking very decorative against the freckled waters of the bay. And, near the water's edge, was Fourah Bay College.

Behind us there was nobody. If we liked we could scale the mountain and arrive, curiously enough, at what was later destined to be the new site of the same college after World War II. Thus, had I but known it, our new home was situated in the direct line joining the old and new homes of West Africa's oldest seat of higher learning.

We were now living in a different world, a happy and united family, sufficient unto ourselves. Our nearest neighbours were a quarter of a mile away. The house itself had a verandah on each of its three floors. On the ground floor a corridor separated the kitchen on the left from the quarters on the right of the two 'boys' who lived in. Other servants, who were gardeners and farm boys, lived in a cottage at one corner of the farm.

In Freetown a 'farm' meant a suburban residential property, usually sited on the forest-clad foothills of the mountains surrounding the city. Our 'farm' was four acres in extent, and mother, who was born in the mountain village of Regent, and was a keen gardener, revelled in turning part of the property into market gardens and orchards, which were both gainful and at the same time brought us children immense delight.

There were now four of us, myself, Arthur, and sisters Phoebe and Irene. Our ages were respectively ten, eight, three and one. Arthur had joined the Government Model School the very next year after me. Our old friends Jabez, and other pals from Kossoh Town, were still as much a part of our lives as ever. But life was more open now, more exciting, with plenty of trees to climb, and all the fruit trees known in Sierra Leone at hand in our own home.

There were dozens of mango trees, including the big cherry, red cherry, rope cherry, and common cherry varieties. There were soursop, sweetsop, black tumbler, finger, oranges, tangerines, limes, damsons, coconut, oil palm, coffee, cocoa pods, and many different varieties of bananas, including those whose skin when ripe was golden, green, or red, and whose shapes were long, short, round or angular. There were plantains, cashew trees, locust trees, paw paw, avocado pears, guavas, different kinds of plums, apples, sugar canes galore, and pineapples.

We made our own sugar (brown), coffee and chocolate, as well as palm oil, coconut oil, and palm-nut oil. We grew ground-nuts, and made ground-nut oil. Sweet potatoes and kassava were staple products grown regularly every year.

When the mango season arrived we would have an invasion of monkeys from the surrounding forest. They would climb the trees silently; but as they ate the mangoes they would start chattering in their excitement, or else fighting, and so give the game away. Then we would dash out and set about them, banging pans and throwing stones, the dogs joining in furiously, and they would scamper away swinging from tree to tree with their long arms and tails.

One day one of them fell down and was immediately pounced upon by Rover, one of the dogs. We rescued it, to find it was little more than a baby. We kept it for a pet and nursed its forelimb which was hurt. Its mother used to come near the house, keeping at a safe distance beyond the fence. At such times the little one would chatter frantically, trying to cut loose from the rope with which it was tethered. Eventually after a week or so father told us to let it go, and immediately it scampered off.

These monkeys were a great pest and destructive to the crops, especially kassava. This they would uproot, and eat the tuber which is the main foodstuff in the plant. However, father was advised to plant what is called 'cotton tree kassava' in the path of the invading animals. From then on the monkeys ceased to eat the plants.

This 'cotton tree' kassava contains a substantial amount of bitter almonds, which contains cyanide, and without special precautions causes sleep and death.

Thus began the period of my life on the farm, amid conditions of perfect happiness and joyous abandon, in sylvan conditions, with dogs and cats, and brothers and sisters, cousins and foster kin, little 'boys' (junior servants) and big 'boys'. For the first time Arthur and I were allowed to linger downstairs in the kitchen in the evenings, and join in the warm after-dinner discussions, and the recounting of 'Nancy stories' and other tales about 'cunny rabbit' (wily Mr Rabbit), Brer Fox, and Mr Leopard. Soon we knew most of the current ones and would join in inventing fresh stories.

During one of these evening sessions I was stung by a scorpion, which may have wandered in from outside. This is the most painful experience I can recall. My heel throbbed painfully for two days.

At holiday times we had rambles into various corners of the farm, which we had not time to explore during school days, and sometimes we crossed into the surrounding fields. There were any amount of trees we could climb without trespassing. They came in handy sometimes when we played hide and seek, especially the biggest mango tree of all, which was just fifty yards from the house. You could shin up there and remain quiet, unless the others had the sense to look up!

At certain times of the year the crickets would be out in strength, and we would go out foraging at night in the garden, guided by their shrill whistles. At our approach they would hop into their holes and lie mute; but it was not difficult to find the hole and start digging cautiously. Soon we would come to the bottom of the tunnel to find Mr Cricket lying still, his wings folded deceptively. We had to grab him firmly or else he would pounce away or kick us with his powerful spiky hind legs. They were very tasty when roasted.

Another roasted dainty was cashew nuts. In addition to the juicy succulent fruit, the stones when roasted exuded a corrosive pungent oil after which the kernel itself became roasted into a fragrant fleshy nut, far richer and tastier than the salted roasted variety sold in packets in foreign cinemas. We used to collect the stones in tins up to hundreds, and sometimes someone would offer us money for them.

Now nature study meant something to me. Early in the morning, even before the cocks crowed, we could hear birds singing in the trees. I could not identify the songs—no one had taught me to—but I saw canaries, and the yellow palm birds which made their nests out of the branches of the oil palm tree. They hung all along these branches, which were otherwise bare. In the evening they came up in great swarms, circling round the palm trees and making quite a lot of happy noise before settling down for the night.

At night we often saw glow-worms darting about with their lights winking in and out. We saw tadpoles in the bed of the stream in our garden. And soon appeared frogs, their throats puffing in and out like bellows, their eyes wandering round cautiously. Now when I heard their croaks at night, I knew what it meant. There were little fishes in that stream.

On the other hand there were snakes. Not many, but certainly not dead in a jar like the one at school. I saw reddish brown ones scarcely distinguishable from the laterite soil; I saw black ones, and green ones. But fortunately our servants were not afraid of them, and soon despatched any that were seen.

There were no paved roads to the farm. Just broad paths for the last half-mile. When later father gave up the hammock for a car, he had to leave it a quarter of a mile away, and walk the rest of the way up. The road was really a succession of outcroppings of pure granite, and we negotiated it more or less like goats on a mountain, stepping from one rock to the other, and occasionally side-stepping a particularly huge boulder.

When we were in a hurry, as we often were, we would take running leaps from one rock to another.

At night we never needed the aid of lanterns. We knew instinctively where each rock was, and exactly when to step to the left or to the right to get around some obstacle, and when to step down from a rock to the lower level of the road and up again. We were never bothered with thoughts of snakes or ants or any untoward things. And we were barefooted. Nothing ever happened to us; we were a part of nature as was our home.

Such then was my life at home. It was perhaps because we lived in such perfect rural, in reality sylvan, conditions, away from the turmoil and jostle of Pownall Street, that I look back on Kossoh Town with such a warmth of affection. Our happiness came from our new home.

But Kossoh Town with St Philip's remained throughout the spiritual centre of our family. There was hardly a day when father did not call at Pownall Street on the way home. It was still the residence of our grandmother, the postal address for our letters. St Philip's always remained our parish church. Often on a Sunday morning the bells of the church would start ringing soon after we had left home, and then Arthur and I would run madly down the hillside, and skim across Malta Street to the church, arriving, with luck, in time to join the procession of choristers moving into the church.

We were indeed more than ever Kossoh Town boys.

Last Year at Model School

AND SO I arrived at Class V, my final year at the Government Model School. From January when the first term started, to the end of the final term in December, the subject which cropped up most frequently in conversation in our group was the Grammar School.

The boys in our class could be divided into three groups. Those for whom this was the last year of schooling ever, those who would be going on to the Wesleyan Methodist Boys' High School, and those who would be going to the C.M.S. Grammar School. These were the two main secondary schools in Freetown for boys, and were maintained respectively by the Methodist and the Church of England (the C.M.S.) Missions. The elementary schools in the city and the surrounding villages were mostly provided by these two missionary bodies. The sister secondary schools were the Wesleyan Methodist Girls' High School, and the Annie Walsh Memorial School.

As for me, my father and all my mother's brothers had attended the Grammar School; and my mother and all my father's sisters had been educated at the Annie Walsh. On both sides we were members of the Church of England, or rather the Sierra Leone Church as it was then called, since it became an autonomous branch of the Anglican Church.

There were no scholarships at all to these secondary schools; nor to any elementary school for that matter. Most of the parents were quite content with the education which was provided in the Model School and other elementary schools. In most cases that was all they could afford.

Many of the companions with whom I sat in Class I had long left. Mohammed Bundukar went the very next year, and by now was an established trader. I sometimes saw him leaning out of one of the second-floor windows of his home, as I passed along Mountain Cut on my way home from school. In the shop on the ground floor his men could be seen handling bushel bags of rice, loads of kola nuts, palm oil, and other commodities in which he dealt.

Horace Chandler and his family had left Freetown. Of the other classmates most of the Mohammedan boys and all of their girls had left. The Mohammedan girls, two from Fourah Bay, one named Abisatu from Foulah Town, stayed with us only through Classes I and II. I never saw any of them again. The Christian girls on the other hand stayed with us much longer, a number proceeding eventually to secondary school.

As to those of us who had survived to Class V, that we had reached this stage at all was a miracle, due almost entirely to the dedicated work of our teachers.

All our education was in English. In fact, everything was in English. I spoke English to my father and the teachers. In church services and on all public occasions English was the medium of expression. All the text-books and indeed all books in the country were in English. The songs at school were British. But all our native life, with its stories, its jokes, dances, songs and thought processes, was in our own native language, Krio. Educated Africans were thus bilingual.

Europeans, Syrians, and other foreigners in our midst, however, persisted in talking down to the natives in what they termed pidgin English. This always sounded very funny to our ears. For this reason, and not from any disrespect, we children used to like to hover out of sight near our principal's office at school. Three Europeans held this office in my time, Mr Holloway, Mr Lean, and Mr Evans.

In those days before the advent of electric fans, the principal sat in his room, with a large broadcloth fan or punka, some four feet wide, suspended from the ceiling above his head by two hinged brackets. A cord passed from this through the lattice above his door to the corridor outside where, all day, a servant sat slowly pulling it to and fro, so generating a current of air above the principal's head.

In this monotonous occupation the inevitable often happened; the man would fall asleep, and the fan cease to move. The principal, deep in whatever he was doing, would gradually realise that the air was stuffy. Looking up he would notice that the fan had stopped. And that was the moment we were waiting for!

'Sori', he shouts with a mighty bellow.

'Sir?', the man leaps up, startled.

'Why for you sleep when you for work?' his irate master storms.

'*A no sleep*, sir.'

'You lie!' thunders principal. The whole idea is to frighten the poor fellow enough to keep him awake for the rest of the day. 'You no lie! you sleep one more time, you be fire, damn quick.'

Actually it would have been easier for everybody if the principal had spoken English. The servant understood it, perfectly, even though he could only express himself in Krio.

* * *

The Government was very proud of the Model School. It was intended as an example to the other primary schools in the Colony. The substantive principal was always a European. But as for us, the pupils, the miracle of our having been educated at all, to the stage when we could soon present ourselves for admission to a secondary school, was due to certain causes despite ourselves.

Regular attendance at school and the meticulous work of our teachers every day, in simple direct language, was bound to have some effect even on the least promising minds. I can picture us now sitting in successive classrooms, looking at the blackboard as the teachers wrote and explained and called for volunteer answers from us. Now one, now another would put up his hand and stand up and answer, mostly correctly, though sometimes wrongly. There was healthy rivalry between us, and we were a fairly well matched group of boys and girls. Discipline was strict. Our teachers saw to it that there was no unnecessary lack of attention, nor failure through playfulness to respond to their demands promptly and efficiently. Failure to do one's homework was punished. So was lateness or poor attendance.

Apart from this we enjoyed going to school. We were young, a new world was opening before us, and every step forward was an adventure. The various subjects we did, nature study, moral instruction, reading, writing, arithmetic, history, geography, scripture, were all new fields, which somehow our teachers made interesting. We were full of zest—surely God's greatest gift to young people.

It was with this same excitement that we received the news that we were to start Latin that year. It was not part of the normal syllabus, but as luck would have it, that year the Rev. W. T. Thomas, M.A., was acting principal, and he felt it would be a great asset to the boys in our class who were due to go to a secondary school. We jumped at it, and I can recall with what thrill I went through the conjugation of my first Latin verb, *amo*, and my first Latin declension of *mensa*.

Mr Thomas also introduced algebra in our group. These two subjects were retained while he remained at the school, until he left to become Inspector of Schools. As a result, during those years, boys moving from the Model School to the Grammar School were usually upgraded one class higher than their predecessors.

There was always keen interest at the placings boys entering that school received, following the entrance examinations.

'I bet we are put in class Intermediate B', one boy said.

'I don't know,' another replied. 'My brother went last year and he was put in Intermediate A.'

'But', the other replied, 'he went from Trinity School. We are doing Latin and algebra.'

I cannot be too grateful to our teachers during those five years at the Model School. One subject which we took at the Cambridge Preliminary Local Examination was arithmetic. By Class V we had completed the whole of the subject, including those complicated problems of men trying to empty or fill a reservoir, or trains running a race, and I have never had to study the subject since.

I must have been benefiting from my schooling by this time, as I found that I looked forward to starting new subjects. I would go through the first few pages as though I was going to devour the book, until my ignorance and inexperience of the subject halted me. Even so I was always reading ahead of my class work. I do not know if this ability to find every new subject interesting is inborn or acquired. Whatever it is, I have retained it throughout my life, and it has been of the greatest help to me.

Indeed so deeply did this contagion of education infect me

at this time that I began to feel ashamed of my handwriting, which was scrawly and unformed. So I bought writing exercise books from T. J. Sawyer's bookshop at Water Street, and spent my holidays practising writing every day.

★   ★   ★

Life certainly was full for us children. On Empire Day, 24 May, all the schools assembled at the Recreation Ground in the western outskirts of the city. Each school assembled separately at its home ground, and marched all the way to the parade ground. The Governor and the Mayor attended and addressed the assembled schoolchildren. Then followed a march past by the schools to music provided by the military band. Finally we dispersed to trek home, in our case at least four miles to the East End.

On Boxing Day the same thing would be repeated, but in a more informal mood. This was the occasion of the Freetown sports, which were held on the same Recreation Ground. The Governor often attended the final sessions of these to present the prizes. The relay races, sprints, and cycle races were great features. These last were for years won by one or other of the three Dworzak brothers. In the centre of the arena was the greasy pole, the most comical competition of all. On the top of this was fixed a leg of ham, and the competitors spent the afternoon doing their best to reach the top; but as the pole was thickly greased, usually it was not until towards the end of the day that somebody eventually managed to reach the prize and grab it.

By now I was roaming freely all over Freetown. On Saturday mornings, and other free days, I would go on errands for mother or some other elder member of the family, sometimes to an aunt who lived in Grassfields, in the West End. One

morning I was walking along the western half of Westmoreland Street, when somebody gripped me firmly by the ear and
pulled me along with him, with the words:

'Where, may I ask, are you going?'

It was the Rev. W. T. Thomas, our headmaster, and I was
forced to hurry along to keep pace with him. He took firm
steps, jutting his neck forward and backward in rhythm like
a hen as each heel came down squarely on the ground. The
shock of this encounter was such that I stammered in Krio,
in reply to his query:

'*Papa sen me, sir*.'

'I beg your pardon?' he queried in terrible tones, and I
corrected myself hurriedly, and replied in English:

'I am on an errand for my father, sir.'

'That's better', he said as we marched along, he still
gripping my ear, but not without kindness. Had he asked me
that same question at that same spot some seven or eight years
later, I could not have answered him without blushing. But
this time my motives were as clear as crystal.

Our church life was developing apace. Arthur and I had
long been stalwart members of St Philip's Sunday School.
Sunday was our most busy day. Three times, morning and
early evening for service, and in the afternoon for Sunday
School, we made the return journey from the farm to Kossoh
Town and St Philip's, and back. Sometimes when one of the
teachers was absent I would be asked to take the Infants class.
This was good training for the time when I became a Sunday
School teacher.

There were many special activities associated with Sunday
School. There were the Pleasant Sunday Afternoon Gather-
ings at which recitations and sacred solos were rendered.
Some distinguished member of the community would take

the chair, supported on the dais by other distinguished ladies and gentlemen. There was the annual special weekday evening concert. This was a secular concert with plays, songs, ditties and sketches, and was usually held on two nights. Then there was the Sunday School party, held on a certain weekday each year when we sat in groups at tables with all sorts of fare, the highlight being Jollof rice, and roast chicken. This roast chicken done the English way, stuffed whole, was a great dainty. The normal African way of preparing the bird is to cut it up and fry it. There were usually gatecrashers at these parties.

But the main event of the Sunday School year was the Sunday School picnic, a spectacular affair. We assembled in church on the appointed day, the girls in white, the boys in black coats, white trousers and straw boaters. At the service prayers were said for the success of the day, the banners were blessed, and we sallied forth with the hymn,

> Onward, Christian soldiers, marching as to war,
> With the Cross of Jesus going on before...

There was always the military or police band in attendance, leading the procession, with our banners, the girls in front, followed by the boys, the older members leading, the youngest in the rear, the teachers walking beside the column, shepherding us.

First we turned westwards, marched to the centre of the city to Government House, and paid our respects to the Governor. Then we turned eastwards again, along Garrison Street, along Kissy Street past the East End Police Station, along Kissy Road, crossing Patton Street from which we had set forth, and so on to Ross Road, and one of the 'farms' there. There we would spend the day as Sunday School picnickers do all the world over. The difference with us was the procession through the city with banners and a band, and the tremendous crush of people accompanying us, including members of our church. The homeward march was even more unrestrained than the outward one; and how we little ones panted and perspired and puffed manfully all the way, our unaccustomed boots pinching our feet!

One year the song ' *Yes, we have no bananas*' had just come out. This was our theme song homeward. The crowd went wild. Even the band went berserk. They played it in the English version, they jazzed it American fashion; they *asiko*'d it to our African dance rhythm. It was gay, it was grand. The cornets, the trumpets, trombones, and bombardons vied with each other; when they tired the bass drums took it up. And when they too were tired our dancing feet kept up the infectious restless rhythm. . . .

*   *   *

And so my time at the Model School drew at last to its close. I was eleven then. At the end of the final term we sat the Cambridge Preliminary Local examination. The papers were to be marked at Cambridge and the result would not reach us until the following spring. But we could not wait for that result to determine our next move. We were all moving on either to the Grammar School or to the High School or to the outside world.

I was considered one of the fortunate ones who would be going to the Grammar School. For had not father made it clear all along the line that the Model School was only the beginning of my education? And had not he and all my uncles attended the Grammar School? So when school broke up, the password among us boys was:

'διώκω.'

This, we had already learnt, was the motto of the Grammar School.

At the terminal examination I came first. I had been first in class throughout the year, and indeed in the aggregate placings at the end of each of my five years at school. My teacher had no doubt that I would be successful at the Cambridge Preliminary examination. The only open point was at what class I should be admitted at the Grammar School. It was indeed with a light heart I said goodbye to my pals.

'Regentonia! διώκω! January!' we shouted to each other as we parted.

But when I went home father called me and said,

'Ageh, you have been expecting to go to the Grammar School next year, have you not?'

'Yes, sir', I said expectantly.

'Well, I am afraid you won't be going this time', he said quietly. I was speechless.

'You see', he continued, 'the Government are going to start a new secondary school at the Model School next year, and I want you to stay with the other boys, to start this new school.'

I could not understand it. I was dumbfounded. It was the collapse of all my hopes and dreams. The Grammar School, the oldest public school in West Africa, part boarding, part day, romantic, full of tradition, the school to which boys came from all over West Africa. What school could possibly be better than that? How could I be better off by going to a different school from that? Besides, this new school had not even started, nobody knew it!

Tears sprang to my eyes. I could not help it. Father let the subject be for the time being, for I was too upset. Later he called me and explained it all carefully.

'Your mother and father know best', he said, and then he asked:

'Have you been happy at the Model School?' Of course I had.

'I know you have,' he said, 'and both your mother and I have been very pleased with the teaching there. It is better than anything you could·have had at the other schools. I am sure that the new secondary school which the Government is starting will be just as good.'

'Yes, sir', I said obediently if rather dully.

'And don't you see,' he continued, 'you are actually making history. You are one of the very few boys to be starting at a school which, God willing, may one day be famous.'

I did not understand him. The possibility he envisaged was too remote for my childish mind to grasp.

'One day you will be able to say "I was the first boy in the Government's secondary school, when it started".' He tried to point out the bright side.

'Yes sir', I said again, and we left it at that.

Not long after I met one of my ex-classmates in town. He was already booked for the Grammar School. He had been admitted to Intermediate B. He it was who had once told me, with great pride, when we were swopping fathers and their positions:

'Oh but *my* dad is a sidesman at the *cathedral*.'

I replied: 'My dad is *pastor's warden* at St Philip's.'

To which he had retorted: 'But that's in *Kossoh Town*!'

This time, assuming that I too was booked for the Grammar School, he asked me:

'What's your new class?'

I hung my head. 'I am not going to the Grammar School', I said.

'Not going to the Grammar School? What do you mean?' he replied, not understanding.

'I am staying on at the Model School. They are going to start a new secondary school.'

'And you are staying for that and not going to the Grammar School?' he asked incredulously.

'Yes, my father says I must stay.'

He shook his head pityingly, murmuring under his breath, as he walked off,

'Kossoh Town boy!'

❦ 19 ❦

Prince of Wales School

SIERRA LEONEANS have a belief that each child brings into this world its own destiny. Its own 'luck', they call it. It is this, they say, which in the long run guides and shapes its future, and not the chance accidents that occur from day to day.

I could not say that I have been a lucky child, in the sense of having anything without having to work hard for it. On the other hand opportunities have opened up often just when the need was greatest. The very year before I started school the Government took a hand in education by establishing the first Government school in Freetown, more than a hundred years after the country had become a British colony.

And here it was again, the very year I was due to start my secondary school career, the same Government for the first time in Freetown decided to start a secondary school. Thus in my short school life I was twice associated with Government projects in education.

This project for a secondary school could have started differently had the Government wished. As had been done in other parts of West Africa, they could have built a proper school, equipped it, brought out teachers from England, and thrown it open to boys and girls from all schools in Freetown and the Peninsula villages.

I do not know what made the Government decide to adopt the method they did in our case, starting from scratch, step by step, with their own boys from the Model School. It could have been the result of mutual mistrust between the Government and the people.

On the one hand the Government might have felt that only boys trained in its Model Primary School were good enough to found a model secondary school under its aegis.

On the other hand it was perhaps too much to expect the cautious Krio parents to send their children to a Government school which was at least unknown if not suspect.

At least by sending them to a recognised Missions Grammar School, if only for a few terms, they would ensure that they received the cachet of an *alumnus*, which in the social and economic conditions of those days was as invaluable as attendance at one of the more famous public schools in England.

The only way in which this difficulty could have been solved would have been to grant a number of scholarships to deserving boys in the colony to the new school. But scholarships were not awarded until almost twenty years afterwards.

So my classmates and I became the little band of unknowns, in this most inauspicious birth of what was destined to be a leading secondary school in Freetown and West Africa. Seven

years later, when the Prince of Wales visited Freetown he
formally baptised this school and conferred on it the honour
of his title. The Prince of Wales School, as it now became
known, moved to its present imposing site at King Tom.

★   ★   ★

The Prince of Wales School, *née* Government Secondary
School, Freetown, started life in a small prison-like room in
the north-east corner of the old manual training building in
the Model School grounds. Actually it was one of the two
halves into which this hipped eastern end of the building was
originally divided. It was the retiring room of Mr Bull, the
instructor. The other was the store room.

In this small, high-vaulted room, we sat facing the door, and
our teacher sat just inside the door, facing us. There were two
windows, both wire-netted, looking northwards into the main
quadrangle and across to the Infants Department.

The manual training classes still went on in the main sec-
tion of the building, but the communicating door between
that hall and this room had been concreted up and a new
door broken into the main eastern wall. Just outside stood
some flamboyant trees which lent a wonderful shade of green,
soothing to our feverish temples and puckered brows.

Did I say feverish temples and puckered brows? It would
be a mistake to think that these were the result of heavy
concentration on serious study. I am afraid that for much of
that first year life for us was a picnic. It was in that classroom
I first learnt to play *ti-ta-to* (noughts and crosses); to throw
ink darts made of blotting paper soaked in ink. We kept a
draughtboard permanently in the class, secreted in our desks,
and played draughts (checkers) regularly during class.

We often wagered how long it would take our teacher to

start nodding. Poor man, for some reason he gave us the impression that he lived a hectic life out of school hours. He always seemed tired and worn out and much in need of sleep. We liked him. But I do not think we learnt much from him.

He was courting one of the teachers in the Infants Department, and often they would walk to school together. Our class assembled half an hour earlier than the main primary school. Secondary schools in the city had earlier hours than primary ones. One morning, as we awaited the arrival of our teacher, we suddenly saw him in the classroom of the Infants teacher. He bent over her from behind, and she lifted her head and neck, like a flower opening when the sun breaks out, and their lips met in a kiss which beat anything I have seen on the cinema screen since. We stared open-mouthed.

We were at the age when we were becoming aware of the anatomy of sex. It was one thing seeing native women and girls carrying water pails and pounding rice in mortar, with naked bouncing breasts. It was an altogether different sensation seeing our own girls, with whom we had grown up from childhood, sprouting suggestive contours on decorously clothed bosoms, and becoming suddenly shy.

About this time a sensation swept through our little class. Until then Krio girls did not engage in public service except as teachers. But now for the first time a group of ex-secondary school girls, from good homes, sat the civil service entrance examination and were accepted. Then came the excitement. They were to appear for a medical examination. Freetown was astir; it was the first time that healthy girls from respectable homes had been expected to bare their bosoms for inspection and examination.

Freetown, which was used to seeing half its female population naked above the waist, was rocked to its foundations

by reports of the physical beauty exposed in these Christian girls, who from early childhood had gone decently clothed.

This was the year when father was away in England, 1919, soon after the end of the War. In his absence, and free from his daily scrutiny of my work at school, my studies suffered terribly. Serious work at school was spasmodic. My classmates and I ran wild, although our spirits were confined to our own select group. At least we maintained our separate identity from the primary departments of the school.

Whereas in the preceding five years I had been a model pupil, within limits, here, although as senior boy in the class I should have set an example, as our teacher himself took little interest in what we were doing I soon wallowed with the rest in empty playfulness.

Yet looking back I can discern a certain amount of gain from this apparent waste. I had been working seriously during the previous years, and as my health was never very strong, it must have benefited from the almost complete absence of mental hurry of that year. I was growing fast and becoming rather lanky. When my father returned from the United Kingdom he expressed concern at my health. So if that year did not promote a healthy mind at least it helped towards a healthy body.

I spent more time on games and often stayed behind with the other boys to play football or cricket at the army pitches on Tower Hill. This we did without boots, and now and again we would play against some of the soldiers, who of course wore boots. I was not good enough to take part in these representative games, but those of our boys who did played very well, and often were so nimble that they caught the soldiers on the wrong foot.

For my part I enjoyed watching and often I was late getting

home. But I had little difficulty in accounting for my movements to my mother, who was not in a position to time me
as efficiently as father did. Father knew our every movement,
and our timetable, both from school and from such extra-
scholastic engagements as choir practice, church classes and
other activities. He knew how long it took us to walk home
from any part of the town and we had to account for our
every movement.

★   ★   ★

I spent three years in this Government Secondary School,
and as it happened those three years could be divided as
follows—the first year when father was abroad and I do not
remember learning much at school; the second year when
father was very much back at home and our class at school
sat the Cambridge Junior Local examination; and the third
year when, instead of proceeding to the Cambridge Senior
Local examination, father made me rest, and start the study
of Greek.

In the three years I was at the school, we had seven teachers,
Mr Wallace, the Rev. Ejesa Osora, Mr Mark, Professor
Orashatuke Faduma, Mr Ladipo Solanke, Mr Williams, and
Mr Sumner. Mr Sumner and Professor Faduma were trained
in America, Mr Osora in England. We never had a European
teacher, although we were administratively under the principal of the Model School proper; that is the elementary school.

With the exception of the Rev. Osora, Mr Sumner and
Professor Faduma, our teachers had all newly graduated from
Fourah Bay College. They came to us for a few months *en
route* for further studies in the United Kingdom, as in the
case of Mr Lapido Solanke, who incidentally was for a time
a sidesman at St Philip's Church; or else they left us to go

to the C.M.S. Grammar School. In later years the current was reversed, and it was more common for teachers to leave the Grammar School for the Prince of Wales School, and the better salary and pension terms of Government service.

All our teachers were University graduates. But such graduates were scarce. The only University College for the whole of West Africa was our Fourah Bay College. It was unusual for as many as six graduates to pass out of the college each year. Of these the majority were men from the other West African countries, and they returned to their homes to teach or enter the ministry or else proceeded to Britain for a profession. Only a handful were Sierra Leoneans and they were not enough to meet the demands for teachers. The Government brought out no one from England to help in actual teaching. Those who came out did so as Inspectors of Schools.

It is unseemly and presumptuous for a pupil to criticise his teachers, even in retrospect; and nothing which I say is meant to reflect in any way on any particular teacher. The fact remains, however, that our progress varied widely with the calibre of our teachers, all of whom must be credited with doing their best for us. However, there were differences in experience, background, and presumably in mental capacity. And we children suffered or benefited from these differences.

As budding secondary scholars we suffered from other disabilities. We had no tradition on which to build. We wore no uniform, no badges, no special cap. There was no house system, no prefect system. There was no annual prize-giving, no parent-teacher association or get-together, although my father managed to keep in touch with what was going on. There were no organised games sessions, no formal competitions, no inter-school matches. In short there was no

PRINCE OF WALES SCHOOL

corporate sense, no organic body to which we could feel we belonged. We were nobody's children. It is not surprising that at the end of the first year many of my classmates left, and the gap had to be filled by merging the remnants with new boys coming up from the Model School that second year.

Also, had we known it, we were face to face with the inherent problem of secondary education as distinct from primary. We were confronted with a new element. Hitherto education had been *imparted* to us. Now we were forced to *think* for ourselves in order to make any headway.

Compare arithmetic with algebra. At the last resort the former can be demonstrated with ten fingers or ten toes, or if necessary by counting piles of stones, or measuring cups of water. But, from the very start, algebra calls for a high degree of imagination, and the capacity for symbolic thought which is the essence of all mathematics. There is all the difference between the ideas and values represented by the letters at the beginning of the alphabet a, b, c, d, and those at the lower end x, y, z. It took me years to realise this, and also that it was the key to Variations. As I see it, the whole method of mathematics as applied to the problems of science consists in reducing the x, y, z's of life to the a, b, c, d's. Logarithms, the Differential and Integral Calculus, and the whole panoply of higher mathematics are aids to bringing about this end.

Take a subject nearer home, geography. In the Primary School we had been concerned with facts, the rivers of the world, the capitals of the world, different countries and counties, their capitals. But now we were forced to think; not only to know the features of certain countries, but why they had those features. For example, why the western seaboards of the British Isles and Norway were rugged and indented with fjords, why the western shores of Britain were rising, while the eastern coastline was in many cases sinking. We had to explain the changes in the English Channel, discuss why Britain is thought to have been part of the European mainland. The changes in the earth's surface in the past, the monsoons, the trade winds, the tides, all these had to be explained. In short we had to be able to use our imagination, and picture in detachment a world rotating on its axis with its satellite moon round the sun, with the resultant physical changes that take place on its surface.

Unfortunately we received little help in this mental travail.

Even before this, we could have been so much helped if we had simply been taken as a class up the slopes of Mount Aureol. From there we could have seen geography spread out below our feet. We could have seen the Bay of Sierra Leone, the capes, the islands, the rivers entering the bay. We could have seen what a basin meant, and many other features of geographical interest. We could have studied cloud formation and movement, how rain was caused, the difference between land and sea breezes. All this would have been a great foundation to our education.

Alas, we had to learn it from books, relying heavily on our imagination.

As for Latin and Greek, there were no introductory talks on life in ancient Rome and Greece, which would have made us realise that the books we were trying to translate were the letters, writings, diaries or compositions of actual people, who had the same peculiarities as people living in our time, just as in the case of a modern poem, novel, or a book of satire or memoirs.

We would have realised that the writings of such authors as Horace, Caesar, Livy, Virgil, Tacitus, Pliny or Juvenal were expressions of different kinds of personalities, just as today some writers have a lyrical turn of mind, some are cynical, some are men of few words favouring the clipped style, others are verbose, some are romantic, some aesthetic, others matter-of-fact.

In the absence of this guide we saw their works as so many pieces of literature which somehow we had to translate and memorise in parts. So many exercises in mental dexterity in fact.

Apart from purely utilitarian considerations, we missed much of the inner meaning and beauty of what we read. Especially in the case of poetry. Our text-books were too often full of scientific minutiae which now in retrospect seem of minor importance compared with the fire and inspiration of the author. Take the famous lines of Virgil:

> Facilis descensus Averni;
> Noctes atque dies patet atri janua Ditis;
> Sed revocare gradum, superasque evadere ad auras,
> Hoc opus, hic labor est.

Even in English they breathe fire and inspiration. But the only thing I remember in the glossary about this passage was a long dissertation on whether Virgil had written *Averni* or *Averno*.

And yet if we imagine Virgil as he was, namely a poet writing in his native language, just like an English poet writing in English, do we think for a moment he would have hesitated to say *Averno* or *Averni* if he thought one or the other expressed more satisfactorily the lilt and inner fire of what he felt? He would most likely have considered it of secondary importance. In any case if he was writing prose he would probably have written *in Avernum*.

Or take the revealing picture of senility conjured up by the four words:

Secunda infantia madidi nasi

the very sound of which nevertheless has all the impact of sheer poetry!

All this was proved beyond doubt in the case of history. I failed in English history when passing Cambridge Junior in 1920 at the Government Secondary School, when I was thirteen. And then the following year the Rev. Ejesa Osora, who had spent many years in England, returned home with an Oxford M.A. degree, and joined the staff of our school among other things as history master. At this time, with the increase in numbers, we had been promoted to a classroom cut off from the opposite or western end of the manual training building, which seemed to suffer progressively as the Government Secondary School grew, and was eventually 'eaten up' in the process.

These history classes were held in the afternoon, and Mr Osora usually made us sit outside in the shade of the trees. In this relaxed atmosphere he made history live for us. The subject was still English history, but he made the kings and queens of England come to life. He described the times in which they lived, and the factors which determined their actions, the choice which they had to make, the dilemmas in which they were placed, and the pull of such factors as economic expansion, the discovery of the New World, religious pressures, and political intrigue. The result was that all of us who took history at the Cambridge Junior that year passed.

*   *   *

This was the year I started the subject of Greek, 1921. I had completed the Cambridge Junior examination the

166

previous year and normally would have gone in for the Senior examination, but father decided that I should rest that year. As a result I had no public examination to take except Greek and supplementary history. I liked Greek even before I started it, just as I had liked the idea of studying Latin. I was soon intrigued by its alphabet and its euphonic sounds. The Greek accents and conjunction also reminded me of our own Krio language in which accentuated pronunciation played such an important part.

Greek was not taught in the school. For this subject I was entirely indebted to Mr Constant Tuboku-Metzger, then a young assistant of my father's at the Water Works, and later one of our leading educationists, who rose to become principal of the Prince of Wales School. He coached me privately in Latin and Greek for both the Junior and the Senior Cambridge Local examinations. His ministrations lasted two and a half years. To him largely I owed the fact that I did uniformly well in these subjects at school and at examinations.

Actually I studied Latin for nine years to 1926, and Greek for seven.

At this time, and through Mr Metzger's influence, I was introduced to American text-books. Unlike the British text-books they concentrated on the essentials and were admirably suited to us, out in Africa, who were dependent on text-books for so much of our education. In England the text-book is an adjunct to the lecture. In Africa the text-book was every-thing.

However, even here I was caught out. With the general slackening of work at school I became lazy, and soon I began to be careless even with the study of Greek, a subject which needs detailed application. Eventually one evening father asked me the usual question: 'What did you do today?' I told him. Then he picked up my Greek Grammar and, idly turning the pages of work I had done recently, he asked me a question here and there. I did not know the answers. He paused a bit, frowning a little, as he often did. Then he turned to the page I had done that very week, that very day.

'Give the principal parts of λαμβάνω,' he asked me.

I tried hard, but I could not give the answer, for the simple reason that I did not know it. He turned another page.

'Give the principal parts of ἐλαύνω.'

Again Ageh was dumb: he could not answer.

Without a word father went out of the room, down the stairs, out into the garden and the night, and came back presently with the freshly cut twigs of the guava tree, whose strokes are more biting by far than an ordinary cane. For about the only time in his life that I can remember he thrashed me with emotion, that is as mother did, without pausing to count the strokes, and only ceasing when he had got over his anger.

That was the last thrashing I can remember receiving from either of my parents. I went up for the examination in Greek and passed. The following year, when I took the Cambridge Senior Local Certificate Examination, I had distinction in the subject, after only two years study of it. Euripides's *Hecuba* was one of the set books. I also studied Hellenistic Greek (the New Testament Greek) and enjoyed it.

To my mind a knowledge of Greek is the crowning mark of a liberal education.

At the end of that year my father said to me:

'Ageh',

'Sir?'

'You will be going to the Grammar School next January.'

I was pleased, but not as excited as I had been at the end of my five years at the Model (Primary) School. I felt I had a stake in the new school which was now building up to a fair size. I had many friends there and in a way I was sorry to leave them.

Many years after, at the end of World War II when I visited the Prince of Wales School, the then principal Mr Davies showed me the old register of the foundation school. The very first name on the list was

Robert B. Cole.

Thus father's prophecy came true and I was Boy Number I on the register of what has since come to be one of West Africa's leading secondary schools.

❦ 20 ❦

C.M.S. Grammar School

AT the Grammar School I settled down from the word Go! I had been happy in the Government Secondary School: after a shaky first year in which we played more than worked, we had settled down into a compact group of scholars, and were fast developing a group consciousness. But, as the spearhead of this new experiment in education we were in an unnatural position, with no group of other boys to look up to, no tradition to follow, no records to inspire or challenge us.

In the Grammar School I was immediately pitchforked into a large and virile group of boys, many of whom were more important than I. Where at the Prince of Wales School I had been Boy Number I, here I was boy number 2946.

The previous December some twenty of us new boys had sat the entrance examination. We were a mixed collection, some from Freetown, two from Waterloo and Hastings, villages some eighteen miles outside Freetown, the others hailing from the sister colonies of Gambia, Gold Coast, and Nigeria. One came from the neighbouring town of Conakry, capital of French Guinea. After the examination I was placed in Form V, the Cambridge Senior Class, and allocated to Primus House. There were four houses then, Primus, Secundus, Tertius and Quartus.

The school building itself was majestic, compared with the single-storied sprawling Model School. Built during the Regency period of British history, and opened in 1845 in what was termed Regent Square, it was partly three-storied, partly four.

In the vast hall extending through almost three-quarters of the first floor were spaced out the various classes. Forms I and II of the Junior School at the south end, the three forms of the Intermediate School along the long eastern side, and Forms III, IV, V and VI on the northern end. For certain subjects the three upper forms used other classrooms near the chapel.

Each teacher stood in front of his class teaching. And believe me there was discipline. When a teacher left his class temporarily for some reason, there was no need to set a prefect to keep order. The slightest misdemeanour from that class brought immediate attention from the nearest teacher.

On a dais along the remaining side bordering Wellington Street sat the principal, Canon Thomas Charles John, M.A., in simple majesty. Later he became Bishop of the Niger Diocese in Nigeria. A portly gentleman with rotund features and friendly eyes, a sonorous voice, and open curly hair, he was one of the great African principals of this famous school, in the direct tradition of Canon Obadiah Moore, the principal in my father's time.

Serious misdemeanours reported to him were often dealt with summarily in front of the whole school. Once a week on a Wednesday afternoon he made a tour of the classes, collecting all the boys who had been late more than once that week. They trailed behind him, a wan little procession, at the end of which marched the biggest boy in the school. They disappeared into another part of the school from which presently sounds of caning could be heard; and soon the boys returned to their classes rubbing their sore bottoms.

The other masters at the school were the Rev. B. L. Thomas, vice-principal, the Rev. Canon E. R. Elliott Spain, Mr Hycy Willson, Mr Frank Marke, Mr E. Morgan, Mr H. B. Williams, Mr E. B. Williams, Mr S. T. Brown, and the next year after I arrived Mr Percy Jones (now Bishop Percy Jones), and Mr Gershon Coker. They were all graduates of Durham University, and had studied at Fourah Bay College.

In this vast human factory each group of boys received earnest tuition from its master. That hall had seen almost a century of this same spectacle. Its floorboards were worn with the march of myriad feet. At regular intervals metal pillars took the weight of the floor above us, where the boarders slept. There were about 130 to 150 pupils at this time, a third of whom were boarders.

The school was run on public-school lines, with its Boarding Department forming the backbone of its tradition. For uniform there was the school cap, the proudest possession of each boy. It was purple with white hoops and the badge featuring a book, a diagonally extended telescope, and the motto διώκω. On Sundays and other formal occasions the uniform was navy blue jacket, white trousers, straw boater and the school ribbon. The ribbon consisted of two purple bands separated by a white one. Fifth-form prefects and sixth-formers alone were entitled to wear the school badge attached to the hat ribbon. This badge was as precious as the colours to an Oxford or Cambridge Blue.

Games were compulsory, each afternoon being reserved for a form or a group of forms. The sports ground was at King Tom, some three-quarters of a mile away, and we usually took the short cut across the creek where the stream of that name entered the sea. We played cricket in the dry season, and football in the rainy season, under the supervision of the appropriate master. Our own form master, Mr Hycy Willson, was Assistant Games Master.

There were competitions for the house shields in cricket, football and athletics, including cross-country running. These were keenly contested, but there was never any methodical practice for them. This year for the cross-country we ran across the creek to King Tom, turned left by King Tom bridge, and up Grassfields, into Pademba Road, turned into Circular Road, and returned by Garrison Street and Westmoreland Road, to school. The whole would be a little over four miles. I never panted so much in all my life.

*   *   *

Each school day started with chapel, after the roll call in class. We marched sedately by class to the chapel, which was situated below the principal's quarters, on the same level as the school hall. The junior boys sat in front, the senior forms at the back. The principal, and the teacher who was on chapel duty each week, sat in the chancel.

The lesson for the day was read by the senior boys in turn. The organist was Mr Frank Marke, one of the masters, and he had one or two assistant organists selected from amongst the boys. Presently I was included in this number. A short address was given each morning. Then we returned to our classes and the first lesson of the day, which more often than not was religious knowledge.

School continued every day in the week including Saturday mornings. There were four terms in the year.

For some reason my father was worried about my health. I do not remember being ill in bed, but he saw Principal John and the latter agreed that I should attend school only in the mornings. When school broke up at twelve I went for lunch to father's office at the Water Works headquarters on Tower Hill half a mile away, before returning home to rest and study at leisure.

I remember having trouble with my eyes. A doctor advised glasses, but there were no eye-specialists and no opticians in the country. When studying at night I used a shade made of freshly cut broad leaves tied over my forehead. This was very soothing, and was changed during the evening as the leaves became dry with the heat.

Also I bathed my eyes in water in which a type of water lily called *ojioro* was grown. A sucker of the plant was placed in an earthenware pot of water. As the roots sprouted down in profusion they were supposed to impart medicinal properties to the water. Night and morning I poured out some of the water and used it, refilling the pot. Incidentally I have never had to wear glasses since.

First thing every term I had to draw up a timetable of homework, based on the timetable of work at school. And from that moment I had to keep rigidly to that timetable.

Each day's work must be done. There was to be no work left over at the end of the week. No study was allowed on Sundays. I must retire not later than ten each night. But there was ample time for rest and relaxation. The main thing was method and regularity in my routine of study. 'Never forget how the tortoise beat the hare', was my father's constant reminder.

In class I had little difficulty in holding my own. The various subjects, Latin, Greek, religious knowledge, geography, history, English language and literature, geometry, algebra, trigonometry, mensuration, shorthand, book-keeping, all were tackled in step with the other boys, although I never quite 'accepted' book-keeping and shorthand.

After six months I was among those who passed the June examination and was promoted to the sixth form. What was more, I headed the pass list and gained my very first scholar-ship, the only scholarship awarded throughout my school career in my native land. It took the form of a reduction of one guinea in the fee of three guineas per term paid by day scholars in the senior school.

That examination was memorable for another reason. I was awarded 100 per cent in one paper in Latin, an achieve-ment which must be bracketed with the zero which I scored in my first-ever examination in geography at the Model School.

We now set to work specifically for the Cambridge Senior examination. New Testament Greek was a special experience. That year the set books were the Gospel according to St Mark and the Epistle to the Galatians. To find the bible, which had been part of my life ever since I could read, now presented in this new language, made Greek all at once take on a new significance for me, while the bible story appeared even more romantic.

We studied Roman history, with special reference to the

period of the bitter wars between the young Roman Republic and the Carthaginians of North Africa. This took us away from present-day English history and the white men in our midst, to days long gone when an equally famous Empire was in the making, and our present masters were unknown. I began to develop a sense of depth in my appreciation of time.

And also the first stirrings of nationalism. To me the Carthaginians were fellow-Africans. I was stirred by the story of how the beaten Hamilcar took his son Hannibal to the temple and made him swear by the God of his fathers to avenge his father's and his country's reverses. The whole saga of Hannibal's taking the long back-door route from Spain over the Pyrenees and the Alps with his mighty army and elephants down to the very gates of Rome fired my imagination to the limit.

Needless to say the villain of the piece for me was the wily Cato, who was largely responsible for wiping Carthage for ever from the face of the earth ('*Delenda est Carthago*').

* * *

My classmates were an interesting lot. There was the boy, some years older than I, who sat behind me, and from time to time when the teacher was absent would hit me on the head with an ebony desk ruler. Goodness knows where he got it from. We all used the yellow flat variety. He probably got it sent as a Christmas present from his native Gold Coast (Ghana) and could think of no more exciting use for it. His middle name was Nwakaku, and his eyes were always red and sultry-looking.

There was another Gold Coaster three rows in front of me in Form IV, a burly gentleman who played centre forward for Tertius House, and used to charge down the centre of the

field like a battering ram, scattering aside the opposition. He had a nephew in the same class; and every now and again he would give the latter a whack with his knuckles if he felt he was lacking in proper respect, and he would mutter through clenched teeth:

'I am your uncle (crack!), your mother's brother (crack!), I pay your fees; treat me with respect (crack!, crack!!).'

As for the masters, the image I have of all of them without exception is of sturdy men with upraised strong right arms carrying canes poised to descend firmly on some boy or other. They all took their duties very seriously. This went back to the days before our time, when education as we know it was not available in West Africa, and often the 'boys' were older than their masters, and the only way the latter could establish their authority was by sheer physical prowess. Especially was this so in the case of the Boarding Department, which accommodated the young and sometimes not so young men from the sister colonies. But, however strict, the masters were nevertheless fair.

Like our mathematics master, Mr S. T. Brown, who claimed to have taught, and thrashed, most of our fathers. No one knew how long he had been at the school. In appearance he seemed a combination of a Yoruba elder and a West Indian sergeant-major. His movements were slow, deliberate and firm, and alone of all the masters he was never seen to smile.

He had a strong sense of fair play. He would set a problem on the board for us to solve. After a while he would get up, stretch his limbs, pick up his cane, saunter to the rear of the class, and working from there to the front would administer a sharp cut to the back of each scholar to hurry him up. After a few weeks he realised that if he started at the back those of us in the front were usually finished by the time he reached us and so missed the caning. He thereupon changed the order of procedure, and from then on would start from the front of the class, working backwards. Thus we all had fair shares.

He would stand us in a bunch in front of the blackboard and start solving a problem. Then, half-way through, he would stop and hand the chalk to one of us and we had to take on from there, while he sat behind us and administered a sharp cut to our calves as we proceeded. These exercises took place in a classroom off the main hall, by the side of the chapel.

The only trouble was that we dared not take our problems in mathematics to him. Anyone who was rash enough to approach him with a problem he could not tackle was liable to find himself planted in front of the blackboard and forced to work out the solution at the point of the cane.

We were proud of our masters. In addition to being our intellectual guides and mentors, they were men of importance in the community. The younger ones especially had great reputations on the football field and at cricket. In those days the Grammar School Old Boys' Association was pre-eminent in Freetown sports.

The *esprit de corps* of the school was solid, morale high. On one occasion following a rather serious disturbance in the boarding department, police-court action was taken against the ringleaders. In an attempt to secure witnesses a number of the junior boys were interviewed by the magistrate. One of the small boys, now a great man, whom we shall call 'John Smith' was asked his name:

'Smith, sir', he replied.

'What is your Christian name?' the magistrate asked.

'Sir?'

'What is your first name? What comes before Smith? What do your friends call you?'

The little fellow thought hard and then replied without the slightest tremor:

'Devil-smile, sir. Devil-smile Smith, sir.'

It was in the same year that the first match between the Grammar School and Bo School was held. The latter is the premier school of the Provinces. These areas were termed 'the Protectorate', 'the hinterland' or 'up the line' (after the railway line) until recently.

Bo School was opened in 1904 by the Government at the town of that name, some 136 miles inland from Freetown, specifically for the sons of ruling chiefs. There was no corresponding school for girls.

This football match between Bo School and the C.M.S. Grammar School was a great event. As far as I know it was the first inter-territorial match between the Colony and the Protectorate. It was played on the neutral Recreation Ground. It was an 'international' affair. Bo School included boys from different chiefdoms and tribes of the Sierra Leone hinterland.

Our team I remember included *Ekeng* (full back, from Calabar, Eastern Nigeria, and belonging to Primus House), *Pratt* (left back, from Freetown, Quartus House), *Asaba* (right half, Sierra Leone and Nigeria, Quartus), *Hesse* (outside left, Accra, Ghana, Tertius), *Addison* (inside left, Accra, Secundus), *Bannerman Bruce* (centre forward, Accra, Primus), *Cookey* (inside right, Bonny, Niger delta, Primus). The Bo team were older, bigger and heavier than our boys.

The Grammar School won. But the match was memorable for what took place between the Bo outside left, whom we shall call Wurie, and our own right half back. After the latter, who was half the size of Wurie, and twice as nimble, had dribbled round him for the *n*th time, Wurie aimed a mighty kick at him, shouting in his exasperation so loud that we at the touchline heard clearly:

'Get to hell out of here! I'm old enough to be your father!' (*Ya, ya! tap am! tap am! a kin bon yu!*)

However it is only fair to add that Bo School beat us in the return match.

21

Interlude

I DEVELOPED rapidly during my first year at the Grammar School. My mental horizon widened. Whereas until then the boys and girls whom I knew were those who lived in my immediate neighbourhood, now my circle of acquaintances rapidly included even youths who lived in some of the very streets along which I had so often passed before.

That I had not been aware of them was largely because obviously we had all gone off each morning to different schools where we had spent the greater part of the day, to return home to our own small groups of associates. Now streams of boys converged from various elementary schools to the Grammar School, bringing with them a rich admixture of types, backgrounds and interests.

And each boy, whatever his background or his position at school, and however brief the period he was privileged to spend at this Alma Mater, would in later life rather die than do anything which might bring dishonour to his old school.

This common sentiment reached fever heat on the anniversary of the school each year on 25 March, when old boys gathered in their hundreds in the school to sing the school song:

> On the Western side of Freetown, upon an ancient site,
> Where a great and massive building stands, a witness to the light,
> There lives a band of comrades true, a band that's ne'er been rent,
> Since the Fathers to the Old Boys gave the glorious name Regent.
> Live for ever! (*stamping of feet*)
> Stand for ever! (*stamp*)
> Live for ever, Grammar school;
> Live for ever! (*stamp*)
> Sundered never! (*stamp*)
> Regentonians true;
> School, school, school, school, school, school!
> God bless our Grammar School! (*a veritable tattoo of stamping feet*).
>
> Oh 'tis not the nerve or sinew, nor learning store alone,
> That the school upon her sons bestow, which stamps them for her own;
> 'Tis manhood's gleam in boyish eyes, steadfast and true and keen,
> A heart that never quakes at fear, a soul that ne'er was mean!
> Live for ever....

Apart from this the new route to school took me through a different part of the city. On the way to the Model School I had skirted the foothills of Mount Aureol, keeping along the southern suburbs. But now on my way to the Grammar School I passed along the principal streets of the city, along Kissy Road, Kissy Street, Westmoreland Street, via either Garrison Street or Wilberforce Street, and so down Bathurst Street to Regentonia. The principal shops lay along this route, and as we hurried off to school in the morning the business community of Freetown would be coming to life.

By some coincidence these shops used to open in a certain sequence, by which we could guess whether we were on time or running late for school; that is if our own consciences had not already warned us.

As a result whereas on the way to the Model School the people I used to see in the streets had all been Africans, excluding the soldiers, now, along this busy thoroughfare, Europeans, Syrians, and Indians were much in evidence. It was the cosmopolitan area of Freetown.

A special train each morning brought the European officials down from the Government residential area at Hill Station, a thousand feet above Freetown, and deposited them at Cotton Tree Station, in the city, collecting them in the afternoon for the return journey home. Hereabouts at the very centre of everything was Fort Thornton, the Governor's residence, right in the path of my shortest route to school.

Among the other secondary schools the only ones with which we had any dealings were our 'sister school', the Annie Walsh Memorial School, and our rivals the Methodist Boys' High School. The boys of the High School had an *esprit de corps* as developed as ours. In those days before the full development of the Prince of Wales School and the birth of the Freetown Secondary School for Girls, the Grammar and the High schools were the Oxford and Cambridge, the Eton and Harrow, the Yale and Harvard, of Freetown, and engendered in the respective boys and their families equal partisan loyalties.

As for the Annie Walsh, it was the practice among the senior boys of our school to write to the senior girls of that school, expressing the warmest sentiments for some particular girl or other, and asking to be accepted as her special friend. To be so accepted was a great honour, full of romantic thrill

for the fortunate boy. The girls of the Annie Walsh were beautiful. They were the cream of our girls, and included also girls
from other parts of West Africa including Liberia. The
effect of this innocent relationship was inspiring, especially
as we were all in the evocative period of our lives. It certainly
improved our letter writing and command of English.

The letters which we wrote were penned with great care
and in lofty imagery. They were often embellished with
Greek and Latin quotations, although interestingly enough
none of the girls' schools did Latin or Greek. But that did not
deter us from baring our souls in the assumed light of the
classical masters. On the other hand the pleasure derived
when the lucky ones received a few short lines from their
chosen goddesses was beyond words. Couched simply
enough, to us the replies were nectar and ambrosia.

In this connection I was in a particularly happy position, in
that the Boarding Department of the Annie Walsh School
attended Evensong at our church of St Philip's in those days.
They attended Matins at Holy Trinity Church, which was
opposite the gates of their school. Presumably the reason
they came to Evensong at St Philip's was because the service
there was earlier, at 5 p.m. compared with 7 p.m. at Holy
Trinity, and also no doubt the longer distance to St Philip's
and back combined worship with much-needed Sunday
afternoon exercise.

After the service they lined up in formation outside the
west door, under the wary eye of Miss Belford, the Senior
House Mistress, and other teachers including Miss Efulabi
Miller, the senior girls at the head of the column, standing
almost opposite the entrance to the churchyard and vestry.
Thus for years, as a chorister, after disrobing, as I made my
way out, I had a vision of these tantalising goddesses. They

always seemed so happy. The corners of their mouths
trembled ever so suspiciously. Of course they were not
unaware of the commotion they caused among the strange
young men who, not being members of our church, yet
always developed religious mania on Sunday afternoons, as
a result of which they made their way from the four corners
of Freetown to St Philip's, Patton Street, Parish Church of
Kossoh Town.

★   ★   ★

And so life at school continued smoothly, when suddenly
a dispute arose in Freetown, starting as a rumour which in
a few days swept the city like a forest fire. The dispute was
simply this. Should the pupils of West Africa be allowed to
sit overseas examinations set by bodies in the United Kingdom, or would they not be better off with local examinations
framed to suit the 'realities of local conditions'?

Our Government Education Department proposed that
the Cambridge and Oxford Senior Local examinations should
be discontinued, on the grounds that they dealt with matters
foreign to African children, and threw too great a strain on
their minds.

It is difficult to appreciate now the depth of feeling which
this proposal roused in the community. Our people strongly
suspected anything which had not the stamp of England, or
which tended to isolate us from anything which Britain had to
offer us. They saw in this, the existing connection between
Sierra Leone and West Africa on the one hand, and Britain
on the other, the only hope for our future. And first on the
list of this connection they placed education, with its over-
whelmingly English context. All this education, including
reading matter and imaginative training, was in English.

The Government pressed its point; the people continued

to resist. The Press took up the matter; and soon, as often happens in these disputes in West Africa, relationships between white and black in the community deteriorated rapidly along racial lines. Officialdom became 'These people'; 'They'.

'These people do not like us. They want to keep us down', complained our elders.

In the end the Government issued an ultimatum. 'We will see what happens at this year's Cambridge Local examination and the matter will be reviewed after that. But in the light of previous years, unless there are definite signs to the contrary, local West African examinations will be introduced', they said. It was hinted that certain subjects were 'unsuitable' and too difficult for us.

It was under this threat that the scholars of Freetown sat the Cambridge Syndicate Local examination that year, December 1922. To be on the safe side only those boys who our teachers were certain would pass were submitted officially by the school, although many of the others also went up privately.

The results were not expected before the end of the following March or the beginning of April. As the time drew near expectation mounted. The sword of Damocles hung over our heads. Bitterness increased. What was going to happen if the Government proceeded with this threat to take away from us this precious Cambridge Local examination? It was generally realised that a local African examination, however difficult or high its standard, could never receive the same recognition in the outside world as a British one. We were a subject people, and our only chance of survival lay in maintaining these contacts with Cambridge and with Oxford, London and Durham.

And then one Saturday morning father came home, and even I could tell that he had great news. Usually he was a very reticent man who kept his emotions under control. But this day, as he mounted the last few steps up the farm, he shouted for me; and as I came towards him wondering what was the matter, he told us the news.

I had passed. The Grammar School had done very well. All the boys entered by the school had passed, two with honours. Freddie Noah had a third-class honours and I first-class honours, the first ever in our history. Freetown was agog that weekend. We were saved.

But there had been those official hints that certain subjects were too difficult for us. So we waited anxiously for the Detailed Reports. When these arrived it was found that of all the thousands of candidates in the Empire, including Britain, the Dominions, India, the West Indies and West Africa, one student, an English boy, had six distinctions. The next highest number of distinctions, four, had been awarded to two candidates, one an Indian, the other myself.

As to the detailed subjects, in my own case, of the eight which I had taken I secured 'credit' in Roman history and drawing, 'very good' in English and mathematics, 'distinction' in Latin, Greek, religious knowledge and geography.

*　　*　　*

Needless to say that year, 1923, was one of the happiest in my life. Instead of my going on to Fourah Bay College, with others of my classmates, father decided that I should stay on at school for another year, in order to give my brain a rest and my body a chance to develop. I was still only fifteen and growing rapidly. It was then he told me how his own health had broken down when he was Head Boy of the school, and he was preparing to proceed to England, to study medicine, following in the steps of his maternal uncle Dr Jacob Vivour Pratt.

He went to Fernando Po for a couple of years' complete rest before his health could be saved. He probably just escaped consumption. It was decided not to risk sending him to face the snows of England. Instead father decided to enter the ministry of the Church. But again his health broke down when he resumed studies in preparation for entering Fourah Bay College. So, regretfully, he decided to give this up, and he went instead to the Technical School to learn engineering, which it was felt was a more practical profession, whose demands were less severely mental.

Within seven years he had graduated from there through the Public Works Department to take over the newly established Freetown Municipal Water Works from the English superintendent engineer who had gone out to inaugurate the venture. But the scare about his own health returned when his son turned out to be similar in build and health. But I was lucky. I had a wonderful father!

I was now the new Head Boy of the Grammar School, and was promptly christened Dux Primus by the other boys. As a sixth former I now wore boots to school; and the first were my father's old pair. He wore size 10. I could get my feet in them only with difficulty, and had corns in all ten toes for the first few months until my feet shrunk by half a size.

It was in this year that I learnt to dance. This too showed father's objectiveness in his dealing with his children. He himself did not dance, nor did mother, nor any of my aunts. In fact nobody in our family did, but when I started taking it up my father raised no objection.

Father always saw to it that we lived a natural life as

children, with plenty of time for play. But otherwise he cut out extraneous activities that might interfere with our regular studies. But now studies with me became secondary, and I attended and even took part in many a concert, both on Sunday afternoons and on week nights.

I remember the first time we went to one such concert at the Wilberforce Memorial Hall by car. After the concert, which started about 8.30 p.m. and finished after eleven, as we motored home eastward along the quiet starlit streets, I shut my eyes in ecstasy, revelling in the sense of motion, when lo! I could smell the moisture in the pure crisp air each time we crossed one of the streams which came down from Mount Aureol under the road, *en route* to the sea!

It was about this time that father presented me with the book 'What a Young Man Ought to Know'. He could not possibly have known how opportune the timing was. A few years before he had given me 'What a Young Boy Ought to Know'.

As a result of this book, with its sensible and noble interpretations of the natural functions and meaning of the sex organs, I was able to see in sex the most wonderful gift we possessed; something to be cherished and honoured.

As a result, when certain of the other boys in class, one in particular, told us the facts of life as seen and practised by them, instead of seeing them as heroes who were devouring forbidden fruit which the rest of us were too timid or too ignorant to grasp, I found their remarks sordid. I felt they were misusing their precious powers; they were fools wasting their substance. Somehow they never made the grade academically.

⋆   ⋆   ⋆

At this time I also learnt to play tennis, and I joined the Scout movement and rapidly rose to become troop leader.

I entered into the various activities of scouting with keenness, and trained for many of the special proficiency badges. Soon I was Head Boy of the school, Head Boy of my house, Head Boy of the Scouts. But whereas I had been inclined to be cocky in my early years, I was now more humble, more natural, and always ready to be helpful to others.

In fact, as a monitor and prefect, my time in the playground was taken up less in keeping order than in helping boys with their problems in mathematics or translations of Latin and Greek passages. They responded by helping me to keep order among the other boys. Surprising as it may seem, I was popular.

Indeed this popularity landed me in serious trouble on one occasion. I was helping to invigilate at a terminal examination one day in the large hall when a boy asked me the answer to a question. Without thinking I told him, and before I realised what was happening the damage was done. Naturally it was reported, and the principal took the very gravest view of the misdemeanour. But for the respect which my father enjoyed, and my hitherto impeccable record at school, I should most probably not have got off with a severe reprimand.

That experience has left a deep impression on me. I always try to remember not to be too impulsive, but alas, often not before it is too late. It has also made me more sympathetic to young people who fall into scrapes, and on more than one occasion I have only been too ready to help all I can when an appeal has reached me.

'There, but for the grace of God....'

*   *   *

So I came to the end of my schooldays. Ten short years, hardly noticeable in the passage of time.

But this was far from being the end of the road for me. I was still only sixteen; and I was twenty-one when eventually I left my native land for further studies abroad. The intervening five years were spent in study in and out of college in Freetown. My father's prediction, in those far-off days when I was pressing him to let me start school, was fully borne out by events. Once I started school the road indeed was destined to be long.

In later years, father, who from being my severest mentor came in time to be a subdued admirer, and who seemed to think much of my ability as an essayist, said to me:

'Ageh, one day when you feel you can do it, I should like you to write an essay on the text, "He that increaseth knowledge increaseth sorrow, and much learning is a weariness of the flesh."'

I have not been able yet to do this. One day, perhaps, I shall attempt it. Who knows? But just now I must bring to an end what after all is a story about a *boy*. As to when I ceased to be a boy, and became instead a young man, I do not know. Is there a time spot for such things? Isn't it rather a progressive affair, an increasing awareness of an altered outlook that has already taken place?

All I know is that one day I woke up to find that that cocoon of a boy with which we started at Kossoh Town was gone. When it happened I do not know. But I am glad to have been that boy.

Printed in the United States
By Bookmasters